SCOPE 21: The Major Biogeochemical Cy
1983, 554 pp

SCOPE 22: Effects of Pollutants at the Ecosystem Level, 1984, 460 pp

SCOPE 23: The Role of Terrestrial Vegetation in the Global Carbon Cycle: Measurement by Remote Sensing, 1984, 272 pp

SCOPE 24: Noise Pollution, 1986, 466 pp

SCOPE 25: Appraisal of Tests to Predict the Environmental Behaviour of Chemical, 1985, 400 pp

SCOPE 26: Methods for Estimating Risks of Chemical Injury: Human and Non-Human Biota, 1985, 712 pp

SCOPE 27: Climate Impact Assessment: Studies of the Interaction of Climate and Society, 1985, 650 pp

SCOPE 28: Environmental Consequences of Nuclear War
Volume I Physical and Atmospheric Effects, 1986, 400 pp
Volume II Ecological and Agricultural Effects, 1985, 563 pp

SCOPE 29: The Greenhouse Effect, Climatic Change, and Ecosystems, 1986, 574 pp

SCOPE 30: Methods for Assessing the Effects of Mixtures of Chemicals, 1987, 928 pp

SCOPE 31: Lead, Mercury, Cadmium and Arsenic in the Environment, 1987, 384 pp

SCOPE 32: Land Transformation in Agriculture, 1987, 552 pp

SCOPE 33: Nitrogen Cycling in Coastal Marine Environment, 1988, 478 pp

SCOPE 34: Practitioner's Handbook on the Modelling of Dynamic Change in Ecosystems, 1988, 196 pp

SCOPE 35: Scales and Global Change: Spatial and Temporal Variability in Biospheric and Geospheric Processes, 1988, 376 pp

SCOPE 36: Acidification in Tropical Countries, 1988, 424 pp

SCOPE 37: Biological Invasions: A Global Perspective, 1989, 528 pp

SCOPE 38: Ecotoxicology and Climate with Special Reference to Hot and Cold Climates, 1989, 432 pp

SCOPE 39: Evolution of the Global Biogeochemical Sulphur Cycle, 1989, 224 pp

SCOPE 40: Methods for Assessing and Reducing Injury from Chemical Accidents, 1989, 320 pp

SCOPE 41: Short-Term Toxicity Tests for Non-Genotoxic Effects, 1990, 353 pp

SCOPE 42: Biogeochemistry of Major World Rivers, 1991, 356 pp

SCOPE 43: Stable Isotopes Natural and Anthropogenic Sulphur in the Environment, 1991

SCOPE 44: Introduction of Genetically Modified Organisms into the Environment, 1990, 224 pp

SCOPE 45: Ecosystem Experiments, 1991

SCOPE 46: Methods for Assessing Exposure of Human and Non-human Biota, 1991

SCOPE 47: Long-term Ecological Research, 1991

Funds to meet SCOPE expenses are provided by contributions from SCOPE Committees, an annual subvention from ICSU (and through ICSU, from UNESCO), an annual subvention from the French Ministère de l'Environment, contracts with UN Bodies, particularly UNEP, and grants from Foundations and industrial enterprises.

SCOPE 45

Ecosystem Experiments

SCOPE 45

Ecosystem Experiments

Edited by

HAROLD A. MOONEY
Department of Biological Sciences, Stanford University, Stanford, California 94305, USA

ERNESTO MEDINA
IVIC, Centro de Ecología y Ciencias Ambientales, Caracas 1020-A, Venezuela

DAVID W. SCHINDLER
Department of Zoology, University of Alberta, Edmonton, Alberta TG6 2E9, Canada

ERNST-DETLEF SCHULZE
Lehrstuhl für Pflanzenökologie, Universität Bayreuth, Bayreuth, D-8580, Germany

BRIAN H. WALKER
CSIRO, Division of Wildlife and Rangelands Research, Lyneham, ACT 2602, Australia

Published on behalf of the Scientific Committee on Problems of the Environment (SCOPE) of the International Council of Scientific Unions (ICSU)

by

JOHN WILEY & SONS
Chichester · New York · Brisbane · Toronto · Singapore

Published by John Wiley & Sons Ltd,
Baffins Lane,
Chichester,
West Sussex PO19 1UD, England

Other Wiley Editorial Offices

John Wiley & Sons, Inc., 605 Third Avenue,
New York, NY 10158-0012, USA

Jacaranda Wiley Ltd, G.P.O. Box 859, Brisbane,
Queensland 4001, Australia

John Wiley & Sons (Canada) Ltd, 22 Worcester Road,
Rexdale, Ontario M9W 1L1, Canada

John Wiley & Sons (SEA) Pte Ltd, 37 Jalan Pemimpin 05-04,
Block B, Union Industrial Building, Singapore 2057

Library of Congress Cataloging-in-Publication Data:

Ecosystem experiments / edited by Harold A. Mooney . . . [et al.].
p. cm.—(SCOPE : 45)
Includes bibliographical references and index.
ISBN 0 471 92926 3 (cloth)
1. Ecology—Congresses. 2. Man–Influence of nature—Congresses.
3. Ecology—Simulation methods—Congresses. 4. Ecology—Research—Congresses. I. Mooney, Harold A. II. International Council of Scientific Unions. Scientific Committee on Problems of the Environment. III. Series: SCOPE report : 45.
QH540.E36 1991
574.5—dc20
91–7558
CIP

British Library Cataloguing-in-Publication data:

Ecosystem experiments.—(Scientific Committee on Problems of the Environment (SCOPE)
I. Mooney, Harold A. II. Medina, Ernesto. III. Series
574.5

ISBN 0 471 92926 3

Typeset in Times 10/12 pt
by Acorn Bookwork, Salisbury, Wiltshire

Printed and bound in Great Britain by Courier International, Tiptree, Essex

International Council of Scientific Unions (ICSU)
Scientific Committee on Problems of the Environment (SCOPE)

SCOPE is one of a number of committees established by the non-governmental group of scientific organizations, the International Council of Scientific Unions (ICSU). The membership of ICSU includes representatives from 75 National Academies of Science, 20 International Unions and 29 other bodies called Associates. To cover multidisciplinary activities which include the interests of several unions, ICSU has established 13 Scientific Committees, of which SCOPE is one. Currently representatives of 35 member countries and 21 Unions, Scientific Committees and Associates participate in the work of SCOPE, which directs particular attention to the needs of developing countries. SCOPE was established in 1969 in response to the environmental concerns emerging at the time: ICSU recognized that many of these concerns required scientific inputs spanning several disciplines and ICSU Unions. SCOPE's first task was to prepare a report on Global Environmental Monitoring (SCOPE 1, 1971) for the UN Stockholm Conference on the Human Environment.

The mandate of SCOPE is to assemble, review, and assess the information available on man-made environmental changes and the effects of these changes on man; to assess and evaluate the methodologies of measurement of environmental parameters; to provide an intelligence service on current research; and by the recruitment of the best available scientific information and constructive thinking to establish itself as a corpus of informed advice for the benefit of centres of fundamental research and of organizations and agencies operationally engaged in studies of the environment.

SCOPE is governed by a General Assembly, which meets every three years. Between such meetings in activities are directed by the Executive Committee.

R. E. Munn
Editor-in-Chief
SCOPE Publications

Executive Secretary: V. Plocq-Fichelet

Secretariat: 51 boulevard de Montmorency
75016 Paris

Contents

PART 2 GROUP REPORTS

Participants in Ecosystem Experiments Workshop

Peter Attiwill
School of Botany, University of Melbourne, Parkville, Victoria 3052, Australia

Suzanne Bayley
Department of Botany, University of Alberta, Edmonton, Alberta T6G 2E9, Canada

Stephen Carpenter
Center for Limnology, University of Wisconsin, Madison, WI 53706, USA

Katherine G. Ewel
Department of Forestry, 118 Newins-Ziegler Hall, University of Florida, Gainesville, FL 32611, USA

Dean Graetz
CSIRO, Division of Wildlife and Rangeland Research, Canberra City, ACT 2601, Australia

Reimer Herrmann
Lehrstuhl für Hydrologie, Universität Bayreuth, Postfach 101251, D-8580 Bayreuth, Germany

Robert Howarth
Section of Ecology and Systematics, Corson Hall, Cornell University, Ithaca, NY 14853, USA

James F. Kitchell
Limnology Laboratory, University of Wisconsin, Madison, WI 53706, USA

Simon A. Levin
Section of Ecology and Systematics, 347 Corson Hall, Cornell University, Ithaca, NY 14853, USA

James MacMahon
Department of Biology, Utah State University, Logan, UT 84322, USA

Ernesto Medina
IVIC, Centro de Ecologia y Ciencias Ambientales, Apartado 21827, Caracas 1020-A, Venezuela

Harold A. Mooney
Department of Biological Sciences, Stanford University, Stanford, CA 94305, USA

Scott Nixon
Graduate School of Oceanography, University of Rhode Island, Narragansett, RI 02882-1197, USA

David Schindler
Department of Zoology, University of Alberta, Edmonton, Alberta T6G 2E9, Canada

E.-D. Schulze
Lehrstuhl für Pflanzenökologie, Universität Bayreuth, Postfach 3008, D-8580 Bayreuth, Germany

C. O. Tamm
Department of Ecology and Environmental Research, Swedish University of Agricultural Sciences, P.O. Box 7072, S-75007 Uppsala, Sweden

Fritz Trillmich
Max-Planck Institut für Verhaltensphysiologie, D-1831 Seewiesen, Germany

B. H. Walker
CSIRO, Division of Wildlife and Rangelands Research, PO Box 84, Lyneham, ACT 2602, Australia

Richard F. Wright
Norwegian Institute for Water Research, Postbox 333, Blindern, Oslo 3, Norway

H. Zwölfer
Lehrstuhl für Tierökologie, Universität Bayreuth, Postfach 3008, D-8580 Bayreuth, Germany

Participants in CO_2 Workshop

L. H. Allen, Jr
USDA/ARS, Building 164, 1FAS-0621, University of Florida, Gainesville, FL 32611, USA

Dennis Baldocchi
Atmospheric Turbulence and Diffusion Division, National Oceanic and Atmospheric Administration, PO Box 2456, Oak Ridge, TN 37831, USA

Fakhri Bazzaz
Department of Organismic and Evolutionary Biology, Harvard University, Cambridge, MA 02138, USA

James I. Burke
Oak Park Research Centre, Carlow, Ireland

Roger Dahlman
Office of Energy Research, DOE/ER-0385, Washington, DC 20545, USA

Ruth DeFries
US SCOPE Committee, Room HA 594, National Academy of Sciences, 2101 Constitution Ave. NW, Washington, DC 20418, USA

Thomas Denmead
CSIRO, Plant Research Division, Canberra, Australia

Bert Drake
Smithsonian Institution, Maryland

George Hendry
Building 318, Brookhaven National Laboratory, Upton, NY 11973, USA

Andrew McLeod
Central Electrical Research Laboratory, Kelvin Avenue, Leatherhead, Surrey KT22 7SE, UK

Jerry Melillo
Marine Biological Laboratory, Ecosystem Center, Woods Hole, MA 02543, USA

H. A. Mooney
Department of Biological Sciences, Stanford University, Stanford, CA 94305, USA

W. C. Oechel
Systems Ecology Research Group, San Diego State University, San Diego, CA 92182, USA

Lou Pitelka
Electric Power Research Institute, PO Box 10412, Palo Alto, CA 94303, USA

Paul G. Risser
Scholes Hall, Room 108, University of New Mexico, Albuquerque, NM 87131, USA

Hugo Rogers
USDA/ARS, PO Box 79, National Soil Dynamics Lab, Auburn, AL 36831, USA

Jelte Rozema
Ecology and Ecotoxicology, Free University of Amsterdam, De Bolelaan 1087, 1081 HV Amsterdam, The Netherlands

B. R. Strain
Department of Botany, Duke University, Durham, NC 27706, USA

Richard F. Wright
Norwegian Institute for Water Research, PO Box 33, Blindern, N-0313 Oslo 3, Norway

Preface

This book incorporates the results of a SCOPE (Scientific Committee of Problems of the Environment) program on ecosystem experiments. This program brought together a group of international scientists at Mitwitz, Germany, to evaluate the state of our knowledge on the response of ecosystems to major perturbations. This group reviewed past work in this area as well as proposing priorities for future research for the study of both terrestrial and aquatic ecosystems utilizing experimental approaches. This meeting was followed by a more technical one, held in Washington, DC, dealing specifically with one of the priority experiments established for terrestrial ecosystems, that of determining the response of whole systems to enriched atmospheric CO_2. The results of this latter workshop are included in this volume. The SCOPE ecosystem experiment program still has in progress considerations for the design of ecosystem experiments for estuarine systems.

This project was initiated to provide guidance to the developing IGBP (International Geosphere Biosphere Program). It is clear that, in order to develop some degree of predictive capacity of the responses of ecosystems to global change, we will have to rely, to some degree at least, on the utilization of the direct experiments on intact ecosystems, as discussed in the introduction and throughout the volume.

This SCOPE project was supported by the A. W. Mellon Foundation and the German Federal Ministry for Research and Technology. The workshop on CO_2 enrichment experiments was aided by the CO_2 Research Program of the United States Department of Energy and the Electric Power Research Institute. Appreciation is expressed to these organizations, as well as to the many scientists who participated in the program.

Introduction

H. A. MOONEY
Department of Biological Sciences, Stanford University, Stanford, California, USA

There is absolutely no question that the biosphere is being altered in a substantial manner by the activities of humans. Changes in the configuration of land surfaces, massive rearrangements of the earth's biota, and an atmosphere that is increasing in concentrations of a number of radiatively active gases are but a few signs indicating the large magnitude of these alterations. Predicting the ecosystem consequences of these changes is a formidable challenge. How will, for example, the increasing concentrations of CO_2 influence ecosystem functioning and distribution? What kinds of tools do we currently have to even approach an answer to such a question?

As a first approach we can look to the historical record. There have been changes in CO_2 concentrations of the atmosphere in the past and accompanying biotic shifts. However, the historical information available is at a fairly coarse resolution, does not provide direct information on system functioning, and does not give clues as to what to expect when atmospheres are not only enriched in CO_2 but also in ozone, sulfur, and nitrogen gases, as we are seeing today. Also the historical rates of change are slower than recent and projected changes in the environment.

A second approach available to us is the use of simulation models that are built for the most part on information on the responses and reactions of pieces of ecosystems—e.g. photosynthesis of leaves in response to CO_2 increase, increase in decomposition due to increased temperature, and so forth. These models are enormously revealing in pointing to the sensitivity of various component processes to environmental changes, as well as indicating where further information is needed in order to understand system interactions. We have learned, however, that there is an enormous 'buffering' capacity as well as compensations within components of systems such that simplistic predictions of responses to change derived from these models may be misleading even in relatively short time periods.

There is a third approach for deriving an understanding of the consequences of change—experimental manipulation of whole ecosystems. Such experiments can tell us how whole systems, as well as their components, will respond to change both in the short and long term, as well as providing direct tests of the models discussed above.

Ecosystem Experiments. Edited by H. A. Mooney *et al.*

The utilization of ecosystem experimentation has been immensely rewarding, yet has received limited application, in part because of difficulty of replication, and the large expense and long duration of these experiments. However, as the need for knowledge of ecosystem response to change becomes increasingly more urgent we must begin to apply this direct method more widely to complement historical and modelling approaches.

The objective of this volume is to explore the potential of ecosystem experimentation as a tool to understanding and predicting more precisely the consequences of our changing biosphere. We take a broad view of the problem by first examining what we have learned from 'natural' experiments as well as large-scale inadvertent ecosystem perturbations induced by human action. Then the results of explicit ecosystem manipulations are reviewed and finally suggestions are made of the kinds of large-scale experiments that might be useful for the future, as well as an analysis of the available tools to do so.

What can we learn from 'natural' experiments? Trillmich's review of the impact of the 1982–83 El Niño on the Galapagos Islands is revealing. Even though there have been no ecosystem studies as such on these islands there have been intensive population studies of representatives of the various trophic levels, in some cases extending back many years prior to the major climatic perturbation that brought 10 times the normal rainfall. Trillmich is able to piece together the interactions of the impact of the perturbation on and among representatives of the various trophic levels. He concludes, though, by noting that without population measurements extending some years prior to the perturbation it would be difficult to interpret the events immediately following the event. Further, the impact will have long-lasting effects that will require continuing observations to fully document and evaluate.

What can we learn from the 'inadvertent' experiments that have resulted principally from the activities of humans on natural systems. A series of papers in this volume address this issue: Medina for the tropical forests, Tamm for temperate forests, and Graetz for desert systems. All take a different approach to the problem. Medina reviews the information on the impacts of deforestation on nutrient, carbon and water cycles, and shows how these manipulations are having profound influences in the short and long term. These analyses provide insights into the mechanisms involved; however, they are not definitive since any particular study generally lacks adequate long-term pre-disturbance measurements as well as controls, and they most often do not view comprehensive system responses. Medina outlines specific system manipulations that would, however, lead to a capacity to predict the consequences of the kinds of modifications that are, and will be, affecting tropical forest ecosystems.

Tamm quite naturally takes a more historical approach to the analysis of the human impacts of disturbance on temperate forests, since such perturbations have been occurring for millennia. He shows how various tools can be utilized to evaluate the effects of these ancient disturbances. He points out, however, that it is often difficult to detect the signal of human-induced disturbance from that of climate change. He calls for a better-developed theory of ecosystem function in order to utilize more fully historical techniques.

Arid systems, because of their relative simplicity and close coupling to rainfall events, offer the possibility of clearly revealing not just the effects of humans on these systems, but the interaction of human systems with unusual climatic occurrences. Graetz is thus able to unravel the interactive result of a severe drought in the Sahel on a nomadic society and the arid ecosystem of which it is a component. It would be difficult to design explicit studies of ecosystems that would include the interactive influences of societies, although such information is highly desirable.

Schulze, Ulrich, and Herrmann take on the difficult task of evaluating the ecosystem response of exposure to chronic pollutants. Schulze and Ulrich show how the effects of acid rain have been progressing for a long time in conjunction with other influences on system functioning. The correlative findings provide the framework for future explicit experiments but in themselves are not totally conclusive. One thing is clear from their analysis—a single atmospheric pollutant can have effects that ramnify through a whole ecosystem, and because of buffering and feedbacks, cause and effect relationships may be very difficult to unravel, without experimentally probing various parts of the network.

Herrmann, examining varied pollutant inputs and transfers within ecosystems, concludes that the unfortunately too common uncontrolled experiments give us but crude information on rates of transfer processes. He proposes a combination field, laboratory, and modelling effort to more fully utilize the information that we have in developing an understanding of the dynamics of the movement of pollutants within ecosystems.

Those contributors to this volume that assessed the inadvertent, uncontrolled experiments that humans have been performing with increasing intensity and frequency all call for the development of a better ecosystem conceptual base in order to evaluate and understand the complex responses that are noted. They think that, in addition to the development of a more robust ecosystem theory and increased model development, small-scale experiments are essential for understanding the functioning and responses of particular parts of the system.

What about explicit whole-system experiments? Schindler makes a powerful case for the utility of such experiments. He demonstrates that issues that were unable to be resolved by the traditional observational and correlative approaches to the study of natural systems were resolved with relatively simple experimental designs. Yet each experiment seems to give some surprising results, underscoring again the poor state of our knowledge of the comparative functioning of ecosystems. Schindler echos the plea of Trillmich for long-term observations in order to understand fully the responses of systems to perturbation, as well as giving many guidelines for whole-system experiments that he has developed over two decades of experience. He urges that the manipulations be designed to yield widely generalizable results, that they be accompanied by intensive comparisons with smaller-scale experimental or observational techniques that have the possibility of being used more widely. Finally, Schindler gives some very explicit experimental approaches that might be utilized to predict the consequences of global change on northern aquatic systems.

Moving from the comparative simplicity of lakes that Schindler considers to the more complex interface of land and water systems, Bayley and Schindler show how fire can be utilized as a tool to test some specific hypotheses about the regulation of nutrient and water transfer in perturbed systems. Pre-disturbance studies, as well as long-term observations subsequent to fire, made it possible to answer the questions that were posed even in these relatively complicated systems.

Both Wright and Attiwill discuss the utilization of watersheds for ecosystem experiments. Attiwill shows how watershed manipulations have been applied first to understanding system water balance, and subsequently to nutrient balance. He, as well as Wright, discuss the problem of suitable references for such manipulations in which there is often but a single manipulated watershed and a reference untreated forested watershed. This brings up an issue which is addressed more fully in subsequent chapters, and one which is the greatest problem to overcome in the full application of ecosystem experimentation: how to get adequate replication in manipulations that are often very large and generally quite expensive, if for no other reason than the length of time that the measurement series must be continued after manipulation in order to get meaningful results. The report of the aquatic group led by Kitchell addresses this issue, and contends that new statistical tools are available that can, as the group state, 'overcome the constraints of statistical tradition'. This is an issue that needs intense consideration since, from all other indications, the experimental approach to ecosystem science is crucial to bringing our understanding of ecosystem functioning to a new level. Wright gives some practical advice on experimental design of watershed manipulations.

Recently ecosystem-level experiments have begun to examine the influence of atmospheric changes on system function. Wright shows the results of whole watershed acidification experiments, and makes the important observation that 'controlled experiments may offer the only realistic approach to quantifying the effects of future changes that have as yet no modern analog'.

For aquatic ecosystem experimentation the lake has been the most utilized experimental unit, and for terrestrial systems the watershed. What about wetlands? Ewel indicates that basins have been useful but that plots can also be utilized in combination with modeling approaches realizing the limitations of interpretations to larger scales. A further useful approach is the employment of constructed hydrologically isolated 'cells' within the wetlands, analogous to the 'mesocosms' discussed for lakes by Schindler. Each approach has its merits and limitations, and their application depends on the question being posed and the resources available.

The issue of the size of the experimental unit is discussed by a number of the contributors, including Schindler, Wright, and Ewel. Schindler discusses at some length the different sorts of interpretations that can be made depending on the experimental unit size. The problem of scaling has received increasing attention during the past few years as ecologists are considering how to make their findings available to scientists who generally work on larger scales than lakes or watersheds, such as atmospheric scientists. There is no doubt that the scaling issue will receive increasing attention in the years ahead, as more and more disciplines interact in the study of the consequences of global change. Levin contributes to this discussion by showing how mathematical techniques can deal with interpreting processes across large scale dimensions.

The various contributions to this volume thus illustrate the enormous difficulties of interpreting ecosystem function, and response to change, by observational and correlative approaches. They further illustrate the rather impressive 'success stories' of explicit ecosystem manipulations. At the same time the considerable difficulties of designing and operating these experiments for the long time periods that are required are made clear. These problems must be addressed and overcome in order to make continued progress in the development of ecosystem science.

There appears to be no lack of good ideas on the kinds of experiments that should be performed, and even for the kinds of experiments that should receive priority attention because of the threat of a changing biosphere. A group under the leadership of Kitchell addressed these priorities for aquatic ecosystem experiments and one led by Walker, terrestrial experiments. The former group proposed experiments on whole-system thermal enhancements, toxic waste remobilization, and system trace gas emissions. Further discussions will examine these kinds of experiments specifically for estuaries. The terrestrial group concluded that whole-system experiments on CO_2 enrichment and modifications of precipitation and thermal regimes are most urgently needed. The working group led by Strain address the techniques available for manipulating CO_2 concentrations at differing levels of organization, from a leaf to whole systems. They show that the technology is available for these experiments; indeed there have already been successful whole-system experiments utilizing enriched atmospheric CO_2 concentrations. However, these experiments have been limited to a couple of ecosystem types only. We have the capability, and the need, to increase the scope of these experiments.

Part 1

CONTRIBUTED PAPERS

1 El Niño in the Galapagos Islands: A Natural Experiment

F. TRILLMICH
Max-Planck-Institut für Verhaltensphysiologie, Abteilung Wickler, D-8131 Seewiesen, Germany

1.1 INTRODUCTION

El Niño events are recurrent disturbances of the ocean–atmosphere system which originate in the tropical Pacific (Barber and Chavez, 1983; Cane, 1983; Philander, 1983; Rasmussen and Wallace, 1983; Graham and White, 1988). These disturbances vary in strength from quite weak—when they are difficult to distinguish from the normal seasonal cycle of changes—to very strong (Quinn *et al.*, 1987), when their effects can be felt almost everywhere within the Pacific basin and influence the ocean atmosphere system worldwide (Wallace, 1985; Cane, 1986). The 1982–83 El Niño was the strongest event recorded in this century. It coincided with a number of long-term studies in the Galapagos so that it is possible to compare the effects of this 'natural experiment' with baseline data.

The effects of El Niño in the eastern tropical Pacific are mostly considered negative to catastrophic. Fisheries, sea bird, and pinniped populations in Peru and Ecuador suffered great losses (Arntz, 1986; Arntz *et al.*, 1985; Barber and Chavez, 1983; Glynn, 1988; Kelly, 1985; Robinson and del Pino, 1985; Trillmich and Limberger, 1985; Wooster and Fluharty, 1985). These events have been documented (Arntz, 1986; Barber and Chavez, 1983, 1986) and the causal chains leading to these effects on the top predators in the marine ecosystem are reasonably well understood. The effects on terrestrial ecosystems are less documented, although they are no less dramatic (Jordán, 1983) and economically at least as important.

The purpose of this chapter is to use some of the available information on the influence of El Niño on terrestrial biota in Galapagos to explore the way in which El Niño changes ecological and evolutionary processes. In particular I want to focus on three potentially very important effects: (1) the influence of the natural El Niño disturbance on the rate of introduction and establishment of new species in the island ecosystem; (2) the selection caused by the disturbance on attributes of Darwin's finches; and (3) examples of long-term effects at the population level.

Ecosystem Experiments. Edited by H. A. Mooney *et al.*

No planned ecosystem studies were made on the Galapagos during the 1982–83 El Niño event. While the event ended five years ago, many of the processes which it set into motion are still strongly felt; also many relevant results have not yet been published and final conclusions may only become possible many years from now. Therefore, the conclusions and interpretations given here must be considered preliminary only.

1.2 BACKGROUND INFORMATION ABOUT THE GALAPAGOS

The Galapagos islands are a group of about seventeen larger islands and about sixty smaller islets and nameless rocks (Fig. 1.1) belonging to Ecuador. They straddle the equator in the Pacific about 1000 km off the coast of South America. Total land area is about 7900 km^2 distributed over an area of about 90 000 km^2 of ocean. The archipelago is volcanic in origin. The oldest islands

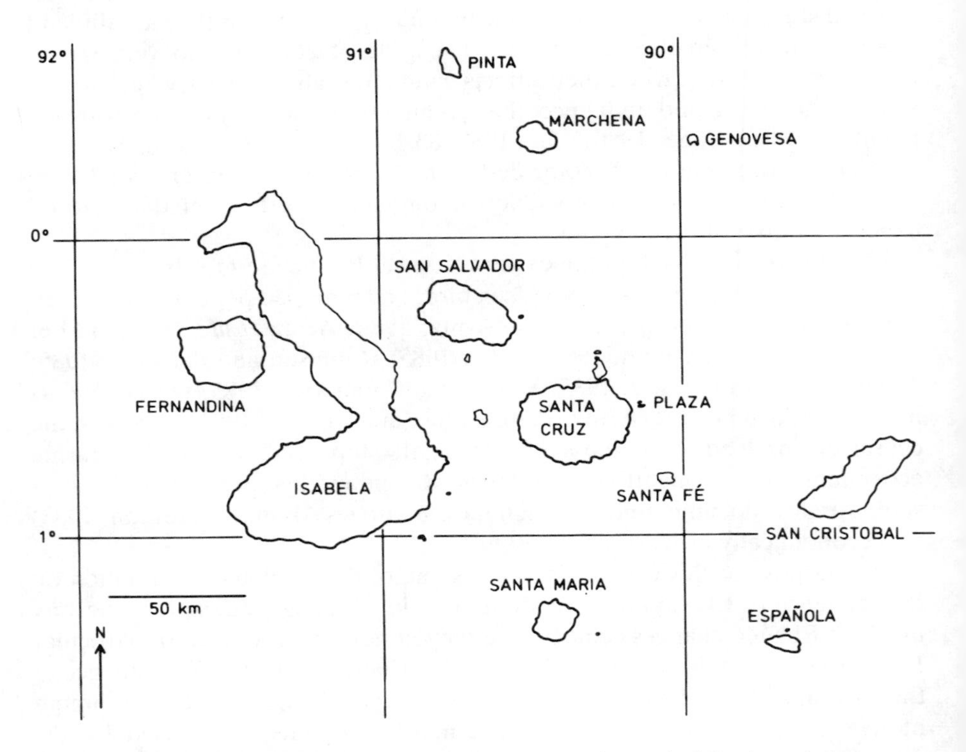

Figure 1.1 Map of the Galapagos islands showing the main islands and site names mentioned in the text.

lie to the east, the younger and highest ones are in the west. Many of the smaller islands are very low and are totally arid, while the younger still active volcanoes in the west reach heights of 1100–1700 m and have extensive humid zones. About 90% of the land area is National Park. Only four islands have settlements (San Cristobal, Floreana, Santa Cruz, and Isabela) with a total of about 10 000 inhabitants, most of whom live in the villages on San Cristobal and Santa Cruz.

1.2.1 THE NATURE OF THE EL NIÑO PHENOMENON

The eastern tropical Pacific is unusually cold among tropical oceans. This is caused by equatorial and coastal upwelling and influx of cold water with the Humboldt current. The system is driven by SE winds which force surface waters away from the South American coast of Peru and Chile leading to local upwelling. In contrast, the surface waters of the western Pacific are the largest pool of very warm water in the world (Cane, 1986). This is maintained by the oceanwide trade winds which drive currents westward along the equator. These water masses warm under the tropical sun and thus bring about the warm water pool in the west. The large gradient of sea surface temperature (SST) between the eastern and western tropical Pacific results in a direct thermal circulation along the equator: cold air flows from the east to the west. The air gets heated and takes up moisture as it flows westward and consequently the air rises in the west. Some of this warm moist air returns to the east where it sinks over the eastern Pacific. This pattern of air flow results in low pressure over the western and high pressure over the eastern Pacific (Wyrtki, 1982).

El Niño is connected with this weather system over the Pacific ocean and in most cases it coincides with significant changes of the 'southern oscillation' the index of which corresponds to the air pressure difference between Tahiti and Darwin, Australia. As explained above, under normal, non-El Niño conditions, pressure is high over Tahiti and low over Darwin. In most cases when this pressure difference is reversed, i.e. when the southern oscillation index sinks to a low level, the consequence is an El Niño. Because of this coupling the phenomenon is often referred to as El Niño–Southern Oscillation (ENSO).

When the trade winds weaken, a normal event at the beginning of each southern summer, the cool Humboldt (Peru) current, its undercurrent and the equatorial subsurface Cromwell countercurrent slow down and upwelling may be diminished (Fig. 1.2). This leads to less influx of cold water, and since the surface heating rate is maintained the surface waters warm. Consequently this season, roughly lasting from December to April, is the warm season as well as the rainy season in Galapagos. When the SE winds gain strength again in April–May normal cool conditions return, and there is frequently an

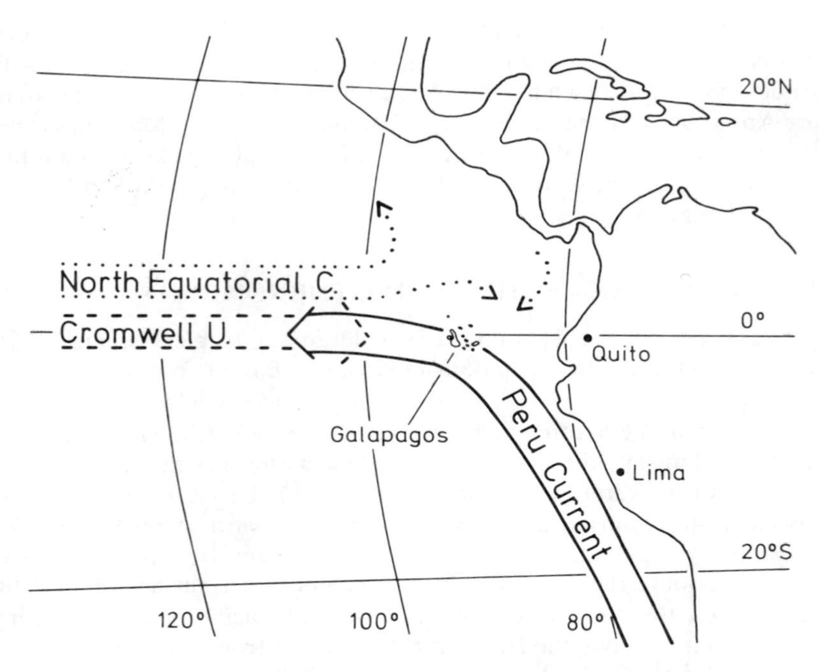

Figure 1.2 Currents influencing the climate of the Galapagos. Subsurface currents are hatched.

inversion layer near the top of the mountains in Galapagos. A very thin cloud cover may form and slight drizzle ('garua') occurs quite frequently.

El Niño typically develops as an amplification of the normal seasonal cycle. However, despite all efforts by meteorologists and oceanographers its occurrence is still unpredictable. Quinn *et al.* (1987) have shown that moderate to very strong El Niño occur at a frequency of about once in 3.8 years (Fig. 1.3), but as stated before the onset, as well as the strength of the event, defeat long-term prediction (but see Barnett *et al.*, 1988). When El Niño occurs, warming begins in January and continues to increase until it peaks in April to July. From July to October the waters cool again and a second weaker warming occurs in the next January. The positive SST anomalies usually range from 2 to 4 °C (Wyrtki, 1982).

The 1982–83 El Niño was unusual in the way it developed, the length of time it persisted, the extent of the geographic area it influenced, and the intensity of the anomaly (Cane, 1986). While most of the preceding El Niño events lasted from January to July of one year (with much less effect in January to March of the next), the 1982–83 El Niño was first noticed in Galapagos in September 1982 and lasted until August 1983. During this whole

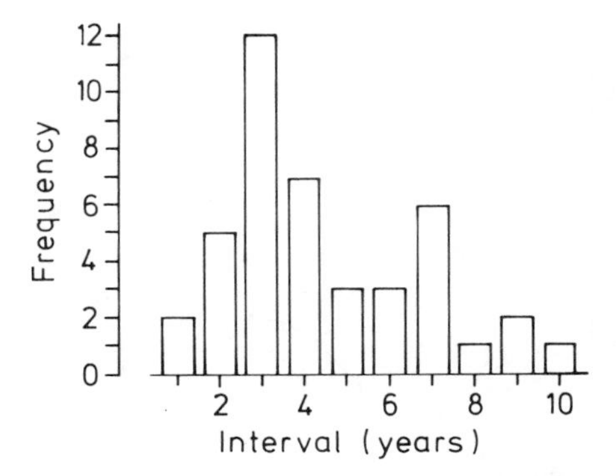

Figure 1.3 Frequency of intervals between El Niño events after Quinn, Neal and Antunez de Mayolo (1987). Only intervals between moderate to very strong events were used in this graph, omitting the less well-defined weak El Niño events.

time, except for a small cooling in February 1983, SST was very much higher than normal with a peak anomaly of 8 °C.

Accompanying this tremendous increase in SST, and most important for terrestrial ecosystems in Galapagos, was torrential rainfall (Fig. 1.4; Robalino, 1985) and an increase in sea level (Wyrtki, 1985) which varied between 20 and 45 cm above normal. Flooding and enormous rivers washed away parts of the road on Santa Cruz, and many buildings were damaged by the floods (various articles in Robinson and del Pino, 1985). Total rainfall during El Niño averaged about 10 times the normal amount. Rainfall peaked initially in December 1982 and January 1983, then decreased slightly in February 1983 and reached a second very high peak in May–June of 1983 (Fig. 1.4).

In August 1983 the El Niño faded away, SSTs returned to normal and the rains ceased. SSTs then continued to decrease and were unusually low in the years 1984 and early 1985. Connected with the low SSTs this period was unusually dry. Between January and April 1985 hardly any rain fell and the islands became extremely dry (Fig. 1.5). The ocean–atmosphere system thus showed a strong rebound effect during 1984–85, producing unusually cool and dry conditions for a period almost as long as the preceding El Niño had been.

1.3 INTRODUCTION AND ESTABLISHMENT OF NEW SPECIES

1.3.1 PLANTS

The Galapagos vegetation can be divided into four main zones depending on the altitudinal humidity gradient. The narrow littoral zone is followed by an

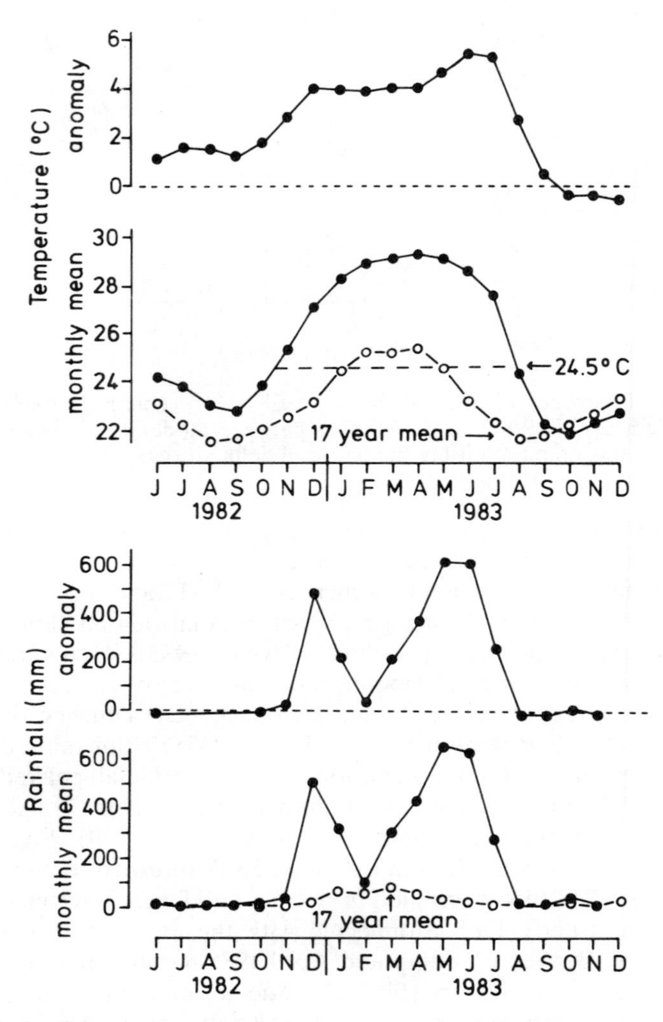

Figure 1.4 Rainfall and temperature and their anomalies during 1982 and 1983 on Santa Cruz Island (after Robalino, 1985).

arid zone which extends roughly to a height of 100 m. The following transition zone gradually merges at about 200 m into a humid zone higher up the mountainside. On the highest islands a secondary dry zone follows above about 500 m, which is mainly covered with ferns and sedges (Wiggins and Porter, 1971). Many of the smaller islands are so low that their vegetation belongs entirely to the arid zone.

The flora of the Galapagos islands consists of about 500 vascular plants, roughly 30–40% of which are endemic. The majority of these plants (about

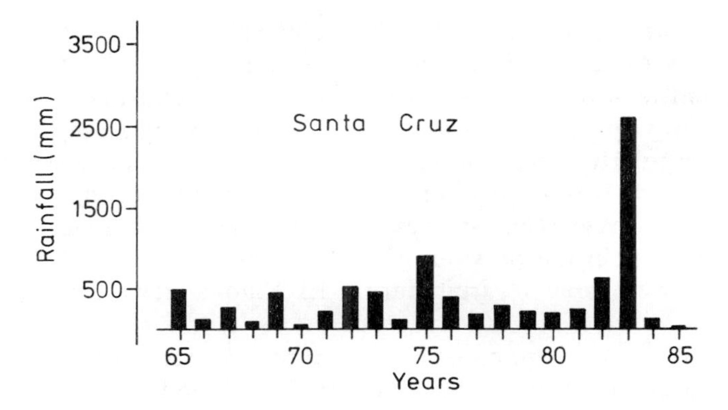

Figure 1.5 Yearly rainfall on Santa Cruz for 1965–85 (after Robalino (1985) and Grant and Grant (1987), changed). Note the extremely low rainfall in 1984 and 1985.

60%) have supposedly been transported to the islands on birds or in their guts, and only about 8% might have arrived by oceanic drift (Porter, 1983). The torrential rains caused by El Niño on the mainland of South America must have flushed a lot of seeds and whole plants from rivers such as the Guayas out to sea, which then drifted along with the currents in the general direction of Galapagos. Drift plants (such as mangroves, *Scaevola plumieri*, *Batis maritima*, *Salicornia fruticosa*, and *Ipomoea pes-caprae*, for example) adapted to life in the mangrove, and littoral vegetation may thus arrive in increasing numbers at the islands during El Niño. Speciation and the number of endemics in the littoral vegetation is, however, quite limited, suggesting that arrival of dissiminules from the mainland is quite frequent and sufficient to prevent genetic differentiation. Therefore, additional arrivals of seeds or other propagules of plants of the littoral zone during El Niño events may not be very important to the ecosystem.

Arriving on an island is, however, only half or perhaps less of the battle to get established on it. Most of the seeds that get washed ashore by oceanic currents, or are carried to the islands by long-distance migrant birds, arrive in a very inhospitable, arid climate. The migratory birds which carry seeds are mostly coastal birds; 20 of 22 species listed as regular migrants to the Galapagos by Porter (1983, after Levèque *et al.*, 1966) are living in the littoral zone. There is a definite advantage, then, if the arriving propagule has the ability to remain dormant until conditions become better. McMullen (1986) showed that 62.5% of the indigenous plants had pronounced seed dormancy (27 species out of 18 families were tested). Thus it seems that many of the plants that became established on the Galapagos were able to wait for quite a while after seed arrival until conditions became suitable for germination. El Niño may help the initial establishment of newcomer plants by providing a mesic habitat in the normally arid zone of the Galapagos. The nine months rain which poured down during the 1982–83 El Niño most likely could break the dormancy of seeds which had arrived earlier, and thus lead to initial establishment of a plant species. Hamann (1985) found that many unusual species of herbs, grasses, sedges, and vines were growing during El Niño in a number of regularly surveyed quadrats (see below), again indicating that a large seed reserve was waiting in the arid zone until moist conditions arrived. About 30% of these plant species had never been recorded in regularly checked quadrats in the previous 17 years (Hamann, 1985). Since most of these plants were able to fruit during El Niño a new seed reserve was deposited.

Since the plants of more mesic habitats cannot survive through dry periods in the arrival site, they either have to produce seeds and return into dormancy *in situ* or else disperse through birds or other animal agents up into the more mesic highlands of the Galapagos. One way in which seeds could be transported into the highlands can be deduced from Cayot's (1985) report on the behavior of giant tortoises (*Geochelone elephantopus*) during the El Niño rains. She observed tortoises on Santa Cruz descending from their normal mesic habitat half-way up the mountain to coastal sites where they fed on the abundant weedy vegetation and fruits until the rains ceased. As the coastal zone dried up the giant tortoises returned to their normal habitat. In this kind of migration they could easily carry seeds or fruits internally or externally up into their mesic habitat. Through this complex interplay of El Niño-related changes, 'stranded' seeds could end up in mesic habitats where a permanent population might become established. Hamann (1984) mentions a few species of drift plants that show very patchy distribution in the Galapagos islands which may have colonized in this way (Table 1.1). For example, *Hippomane mancinella* is a typical oceanic drift plant which has colonized both coastal and inland habitats in the Galapagos. *Hippomane* fruits may easily have been transported inland by giant tortoises. Tortoises were observed to eat *Hippomane* fruits, which then, often after several days, get passed almost unchanged in appearance (Hamann, 1984).

A second category of plants with drifting propagules occurs only inland. This category includes such species as *Mucuna rostrata*, *Dioclea reflexa*, *Sapindus saponaria*, *Stictocardia tiliifolia*, and *Ipomoea alba*. In the Galapagos these plants have restricted and spot-like distributions. This distribution pattern is difficult to explain in relation to the plants' effective dispersal mechanism. This may reflect that although they are well adapted for long-distance dispersal through oceanic drift, they could be less well adapted to establish in an arid coastal region such as that on Galapagos, because they require mesic habitat. Thus, once arrived on the beach, they would have to get a rare opportunity to cross the barrier of dry land to become established. This could limit their success as colonizers.

Hamann (1984) concludes: 'The present spot-like distribution patterns of

Table 1.1 Some drift seed plants of the Galapagos; this lists only some of the more conspicuous species (after Hamann, 1984)

Species and family	Propagule	Distribution
Plants of coastal and inland habitats		
Caesalpinia bonduc (Leguminosae)	Seed	++
Hibiscus tiliaceus (Malvaceae)	Seed	++
Hippomane mancinella (Euphorbiacea)	Fruit	++
Plants of mainly inland habitats		
Sapindus saponaria (Sapindaceae)	Seed/fruit	++
Stictocardia tiliifolia (Convolvulaceae)	Seed	++
Ipomoea alba (Convolvulaceae)	Seed	++
Mucuna rostrata (Leguminosae)	Seed	+
Dioclea reflexa (Leguminosae)	Seed	+

+, Spot-like distribution; ++, locally abundant.

some notorious drift seed plants in the Galapagos suggest that successful colonization by these species is a rare phenomenon. However, extreme environmental conditions such as those in a Niño year could be a decisive factor.'

1.3.2 ANIMALS

One animal population has increased dramatically during the 1982–83 El Niño. The smooth-billed Ani (*Crotophaga ani*) was presumably introduced in the early 1960s to the Galapagos (D. Rosenberg in *Carta Informativa*, No. 21, August 1987, from the Estación Charles Darwin and Servicio Parque Nacional, Galapagos). It remained rare in the 1960s and 1970s, to the extent that it was classified as a straggler by Harris (1974). During El Niño, however, these birds, just as most of the endemic land birds (see below for an example from the Darwin's finches) reproduced almost continuously, presumably because prey abundance increased drastically. This led to a population explosion which was not stopped by the drought period following El Niño. The population of Anis on Santa Cruz is presently estimated at about 5000 individuals (D. Rosenberg, *loc. cit.*). From there the birds have begun to disperse to nearby islands such as Santa Fé and Pinzon. As this bird is a nest predator of smaller birds its population increase and dispersal to other islands may have serious long-term effects on the abundance and species composition of the avifauna of the islands on which it becomes numerous.

The examples from plants and birds demonstrate that populations which initially just barely escape extinction in their new habitat can multiply rapidly if El Niño-related rains increase the suitability of the habitat for a short while.

This increases the chances of plants to find a permanently suitable habitat, and allows populations to escape from dangerously low levels to a size where the probability of extinction is low even after the disturbance is over.

1.4 THE EFFECTS OF EL NIÑO ON DARWIN'S FINCHES

Darwin's finches, and among those mostly two species of ground finches (the medium ground finch *Geospiza fortis* and the Cactus finch *G. scandens*) have been studied for up to 12 years by Grant and his co-workers (Grant, 1986; Gibbs and Grant, 1987a–c). Darwin's finches are small passerine birds that are famous for their tremendous adaptive radiation on the Galapagos. They breed during the wet season. The ground finches live mostly on seeds, but also feed on pollen and nectar produced by the opuntias. During the dry season the birds live almost exclusively on small seeds that have been deposited during the previous wet season. Bigger birds were found to be most efficient in cracking larger seeds, since they have stronger beaks which allow them to exert more force. Most of the data which I want to review here come from ground finch populations studied on Daphne Major, a *ca*40 hectare volcanic islet situated in the center of the archipelago. Ground finch populations there were color-banded and studied every year during the breeding season.

During El Niño the unusually strong plant growth produced a large seed crop, which by June 1983 was 11 times greater than in 1982 (Fig. 1.6). Up to 80% of these seeds were small, that is, of the size which is easily cracked by the ground finches. So large was the seed crop produced during 1982–83 that small seeds remained very abundant for two years after the El Niño.

This situation of high food abundance led to intensive breeding during El Niño. Birds bred almost continously for nine months. Medium ground finches produced six times as many clutches as in other 'normal' years, and produced 3.5 times as many young. This led to a dramatic change in population structure (Fig. 1.7) which—given the longevity of these finches—will persist for quite some time. From a relatively even age structure with no individuals less than a year of age in January 1982 the populations changed to 75% young birds in January 1983. At the same time this enormous production of young led to a doubling of the population. The increased level of total seed biomass (Fig. 1.6) maintained this population at this high level even two years after the onset of the El Niño event. Thus the carrying capacity of the island had been increased for such a long time through one nine-month period of unusually heavy rain.

Another important effect of the increase in total seed biomass was the establishment of a breeding population of the large ground finch (*Geospiza magnirostris*) on the island during El Niño. Migrant birds had been recorded

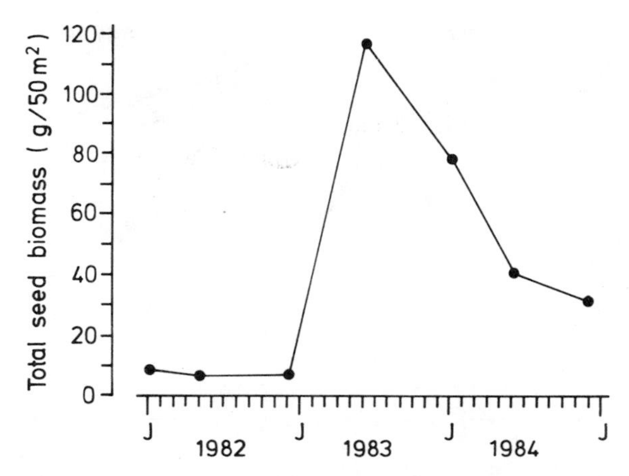

Figure 1.6 Changes in total seed biomass (g/50 m^2) during the El Niño period (after Gibbs and Grant, 1987a).

on the island before, but none had bred there for nine years. This shows that the species composition of the *Geospiza* community can be dynamic, with the presence or absence of rare breeding species depending on the occurrence of rare events such as El Niño or prolonged droughts.

Boag and Grant (1981) had shown that during periods of drought larger *Geospiza* individuals survived better because they were better able to live on remaining large seeds. Since the attributes measured by them (beak dimensions, body size) were shown to be highly heritable (Boag and Grant, 1978) this means that there was strong selection against small body size. During El Niño and the period of food abundance following it selection on body size was opposite to that under food shortage (Gibbs and Grant, 1987b). In 1984–85

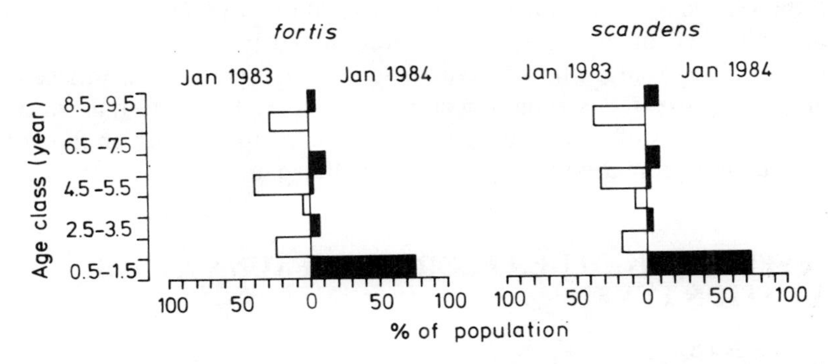

Figure 1.7 Age structure in *Geospiza fortis* and *G. scandens* populations in January 1983 (basically still pre-El Niño) and January 1984 (after the end of El Niño) (after Gibbs and Grant, 1987a).

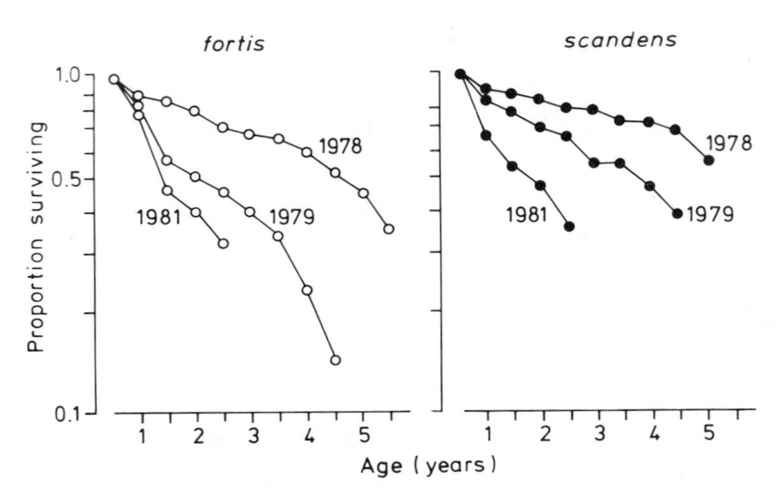

Figure 1.8 Survivorship curves for cohorts of *Geospiza fortis* and *G. scandens* (after Gibbs and Grant, 1987c).

small birds survived significantly better than large birds, and the selection differentials and selection gradients found were mostly acting on size and bill dimensions. This means that rare climatic perturbations maintain these natural populations away from an equilibrium state demographically, phenotypically and genetically.

I want to mention one last effect on the demography of these finches which seems particularly noteworthy. When Gibbs and Grant (1987c) looked at the mortality of different cohorts of the medium ground finch and the cactus finch they found them to vary tremendously (Fig. 1.8). Apparently the first 1.5 years in the life of a cohort determine the mortality rate of that cohort for the rest of their lives. Since the mean annual survival rate of these finches is about 0.7, the influence of environmental conditions early in life can be carried through for many years. The unpredictable environment in which these birds live may have exerted strong selection towards a long lifespan since an individual's chances of successful reproduction depend strongly on the probability of encountering a good year during its lifespan.

1.5 EFFECTS ON OTHER FLORA AND FAUNA AT THE POPULATION LEVEL

1.5.1 PLANTS

Effects, surprisingly similar to those documented for Darwin's finches, have also been found in a number of populations of even more long-lived organisms. A number of mostly Danish botanists have carried out investigations of floral changes since 1966 by means of permanent quadrats representing various vegetation types. Previously collected data allow one to discern between vegetational changes occurring as a normal part of succession or normal fluctuations and those changes caused by El Niño (Hamann, 1985).

Investigations in the arid zone of Santa Cruz island (Fig. 1.1) show that the vegetation was quite stable since 1966. The main woody species in these quadrats were *Opuntia echios*, *Jasminocereus thouarsii*, *Bursera graveolens*, *Croton scouleri*, *Scutia pauciflora*, and *Acacia* sp. The most striking immediate change during El Niño was the tremendous growth of several species of vines. The heavy load of vines, combined with the weight of the fully succulent leaves and waterlogged root system, exceeded the stability of the tree cacti and many of them fell. Over the period of El Niño about 30% of the *Opuntia* cacti died in two quadrats on Santa Cruz and no new plants were able to get established, presumably because they were outcompeted by many grasses and weedy plants that were germinating rapidly.

In the higher parts of Santa Cruz the endemic *Scalesia* tree (*Scalesia pedunculata*) is the main constituent of the forest. These trees have an estimated lifespan of 10–15 years. In quadrats in the forest zone many *Scalesia* trees were dying. Only two of 17 trees were still alive in August 1983, after El Niño. The trees had apparently died through waterlogging of the trunk and flooding of the soil, so that the root system had died. In 1985, after the drought period subsequent to El Niño, 5500 saplings were estimated to grow in the same quadrat. Only 61 of these were still alive in July 1987, when they had reached a height of about 5 m (Lawesson, 1988). Hamann (1979) suggested that *Scalesia* trees are pioneer or early successional trees that are able to persist as dominant tree species through rapid invasion of forest gaps, fast growth, early maturity, and effective short-distance dispersal. Contributing to this success is the fact that most late successional tree species are absent in the Galapagos flora. The effect of this massive die-off of the *Scalesia* was to create a new forest of almost uniform age structure after El Niño. Similar effects have been observed for stand of *S. cordata* and *Miconia robinsoniana* (Lawesson, 1988).

1.5.2 THE LAND IGUANAS OF S-PLAZA

Snell and Snell (personal communication) studied the population of land iguanas (*Conolophus subcristatus*) on S-Plaza, an islet of about 11.7 ha. Their study covered the whole El Niño event. The qualitative results of this study are so interesting that they will be mentioned here although they are not yet published in detail.

Land iguanas on Plazas dig burrows and lay their eggs in January and February, and the young hatch in May to June after about four months of incubation. During El Niño the unusually large rainfall led to massive growth of the vegetation providing the iguanas with plentiful food. This was reflected in excellent body condition of all animals in the population and consequent massive, successful reproduction of the iguanas on Plazas in 1983. Hatching success was high and hatchlings were unusually heavy. High water potential of the soil has been shown (Snell and Tracy, 1985) to result in larger hatchlings. The unusually big El Niño hatchlings emerged into a situation of plentiful food, resulting in high growth rate and low mortality of juveniles. Thus population size rapidly increased during the period of El Niño.

At the same time the Opuntia (*Opuntia echios*), which is one major food item of iguanas during dry seasons, became water-logged and 50% of the trees fell. Usually leaves of fallen trees will resprout and vegetative reproduction can occur. However, under the conditions of high iguana population density after El Niño this did not happen. The reasons are two-fold: immediately after El Niño a period of intense drought followed which reduced the amount of edible vegetation on the low-lying Plaza island very quickly. This secondarily led to preferential feeding of the iguanas on the fallen Opuntias and their resprouting leaves. The iguanas ate all regrowth and reduced Opuntia density to 50% of its former value. This process has been continuing ever since, to the extent that Opuntia reproduction through fruit cannot take place any more. That the iguanas were indeed responsible for this partial destruction of their food base is proven by a natural control. N-Plaza lies only about 50 m north of S-Plaza, has the same physical properties, is covered with the same type of vegetation, but devoid of land iguanas. On this island Opuntias also fell during El Niño, but resprouted normally, resulting in a fast recovery of the local Opuntia population.

With the partial demise of the Opuntias on S-Plaza the staple food of iguanas during the dry season has become very scarce, and as more and more of the old Opuntia trees fall and are eaten by the almost starving iguana population the resource base is increasingly vanishing. This has led to reduced reproduction, increased mortality among adults and almost 100% mortality of juveniles in their first year. In 1988 many females that reproduced were so spent that they did not recover from their reproductive effort and died. Presumably, this population will continue to decrease until the *Opuntias* finally get a chance to regrow, so that the resource base of the iguana population recovers. When this will happen, and through how serious a bottleneck the population will go in between, is presently impossible to say. But given the initial relatively small size of the iguana population on S-Plaza, with 226 adult females only, and 102 adult males (Snell and Christian, 1985), it is conceivable that effective population size could drop well below levels where random drift and founder effects could have a strong impact on the evolutionary trajectory of this population.

Similar dramatic effects of El Niño have been observed in other large vertebrates that depend for their food on the marine environment. Here the effects mirrored the ones found in terrestrial organisms: the period of El Niño was disastrous while populations recovered rapidly during the rebound period.

1.5.3 THE MARINE IGUANAS OF SANTA FÉ

Marine iguanas (*Amblyrhynchus cristatus*) on Santa Fé lay their eggs during January/February and the young hatch in April/May. This population suffered from the immigration of indigestible brown algae during El Niño (Laurie, 1988) and decreased by 70% during the disturbance (Laurie, 1988). The 1982 hatchlings were almost completely exterminated, and the cohorts of 1981, 1980, and 1979 suffered progressively lower mortality. Although breeding was normal in January/February 1983 almost no breeding took place in the 1983–84 season, after the El Niño. Apparently females still had enough reserves for egg production early in the El Niño event, while after the event had ended they were unable to accumulate new reserves for reproduction during the few normal months (approximately from August to December 1983) prior to the breeding season. Due to fast recovery of the normal algal flora and low intraspecific competition for food in 1984, the growth and reproduction of the survivors was very high during 1984 and in early 1985. The observed drastic decrease of this population, coupled with an equally dramatic change in the age structure, is going to remain noticeable for at least a decade after the event (Laurie, 1988).

1.5.4 THE GALAPAGOS FUR SEAL

A similar effect has been shown for the Galapagos fur seal (*Arctocephalus galapagoensis*) (Trillmich and Limberger, 1985). This species is quite dimorphic. Females weigh about 30 kg, territorial males about 65–70 kg. During El Niño high adult mortality (30% in females and about 100% in reproductive males) and extreme juvenile mortality (complete loss of the 1982, 1981, and 1980 cohort) was coupled with hardly any production of young in 1983, immediately after the event.

What was most surprising in this species was the almost total mortality of males that had been territorial in the 1982 El Niño while females survived much better. Territorial males fast during their two to six weeks on territory. The males that did so in 1982 returned to an ocean that provided hardly any food, and they starved to death. Females never fast for so long and their absolute food requirements are about half of those of fully grown males. Presumably due to this difference in foraging regime and food needs, females survived in El Niño much better than territorial males, even though almost all of the surviving females were unable to produce a pup in the 1983 reproductive season (Trillmich and Limberger, 1985) showing that they could not accrue sufficient resources for successful reproduction late in the El Niño. These data show that in sexually dimorphic species the sexes can react quite differently to the effects of an environmental disturbance, as if they belonged to different species.

1.6 CONCLUSIONS

The observations on Galapagos during El Niño show that one has to be aware of previous disturbances to an ecosystem when beginning an experiment. Many of the effects observed in the Galapagos were so long-term that only a study covering several years before the onset of an experiment would have allowed valid interpretation of the experimental period. To assume that a system under investigation is in equilibrium when an experiment is started seems quite unwarranted. This is clearly borne out by the examples in this chapter: 17 year seed dormancy, long-term effects of selection on finches and long-term after-effects on the age of structure of plant, reptile, and mammal populations are difficult to detect at the beginning of a study, but potentially have a great influence on later interpretation of the data. Furthermore, an experimental change in an ecosystem may change the conditions for immigration/emigration of organisms, and in this way change the composition of the system. Last but not least, El Niño demonstrated quite convincingly that variability in an ecosystem is an important component of the system. If such variance creates occasional bottlenecks for populations it may produce conditions for (more rapid) evolution of the component species. Experimentally altering conditions in an ecosystem and observing the response of the system and its components short term, may thus lead to very short-term predictions only if it is overlooked that important components of the ecosystem under study may change over a few generations due to directional selection under the altered conditions.

ACKNOWLEDGEMENTS

I would like to thank the Galapagos National Park Service under its Intendentes M. Cifuentes, F. Cepeda, and U. Ochoa for permission to work in the Galapagos, and to thank them and the directors of the Charles Darwin Station, C. MacFarland, H. Hoeck, F. Köster and G. Reck, for their continuous support of my own work in the Galapagos, as well as for many interesting discussions. Ole Hamann, Henning Adsersen, and Jonas Lawesson discussed botanical questions with me and kindly provided published and unpublished material. I greatly appreciate their help and patience. A. Laurie and H. and H. Snell filled me in with exciting information on iguana life during El Niño and allowed me to use their unpublished data. Max-Planck Gesellschaft, through W. Wickler's permanent help, encouragement and support, made my work in Galapagos possible. This is contribution No. 433 of the Charles Darwin Research Station.

REFERENCES

Arntz, W.E. (1986) The two faces of El Niño. *Meeresforsch.* **31**, 1–46.

Arntz, W.E., Landa, A. and Tarazona, J. (Eds) (1985) El fenómeno 'El Niño' y su impacto en la fauna marina. *Bol. Inst. Mar Peru-Callao* (special issue).

Barber, R.T. and Chavez, F.P. (1983) Biological consequences of El Niño. *Science* **222**, 1203–10.

Barber, R.T. and Chavez, F.P. (1986) Ocean variability in relation to living resources during the 1982–83 El Niño. *Nature* **319**, 279–85.

Barnett, T., Graham, N., Cane, M., Zebiak, S., Dolan, S., O'Brien, J. and Legler, D. (1988) On the prediction of El Niño of 1986–1987. *Science* **241**, 192–6.

Boag, P.T. and Grant, P.R. (1978) Heritability of external morphology in Darwin's finches. *Nature* **274**, 793–4.

Boag, P.T. and Grant, P.R. (1981) Intense natural selection in a population of Darwin's finches (Geospizinae) in the Galapagos. *Science* **214**, 82–5.

Cane, M.A. (1983) Oceanographic events during El Niño. *Science* **222**, 1189–95.

Cane, M.A. (1986) El Niño. *Ann. Rev. Earth Planet Sci.* **14**, 43–70.

Cayot, L.J. (1985) Effects of El Niño on giant tortoises and their environment. In: Robinson, G. and del Pino, E.M. (Eds), *El Niño in the Galapagos Islands: The 1982–1983 Event*, Charles Darwin Foundation for the Galapagos Islands, Quito, Ecuador, pp. 363–98.

Chavez, F. (1987) The annual cycle of SST along the coast of Peru. *Trop. Ocean-Atmos. Newsl.* **37**, 4–6.

Gibbs, H.L. and Grant, P.R. (1987a) Ecological consequences of an exceptionally strong El Niño event on Darwin's finches. *Ecology* **68**, 1735–46.

Gibbs, H.L. and Grant, P.R. (1987b) Oscillating selection on Darwin's finches. *Nature* **327**, 511–13.

Gibbs, H.L. and Grant, P.R. (1987c) Adult survivorship in Darwin's ground finch (*Geospiza*) populations in a variable environment. *J. Anim. Ecol.* **56**, 797–813.

Glynn, P.W. (1988) El Niño-Southern Oscillation 1982–83: nearshore population, community and ecosystem responses. *Ann. Rev. Ecol. Syst.* **19**, 1–40.

Graham, N.E. and White, W.B. (1988) El Niño cycle: a natural oscillator of the Pacific ocean–atmosphere system. *Science* **240**, 1293–302.

Grant, P.R. (1986) *Ecology and Evolution of Darwin's Finches*, Princeton University Press, Princeton, NJ.

Grant, P.R. and Grant, B.R. (1987) The extraordinary El Niño event of 1982–83: effects on Darwin's finches on Isla Genovesa, Galapagos. *Oikos* **49**, 55–66.

Hamann, O. (1979) Dynamics of a stand of *Scalesia pendunculata* Hooker fil., Santa Cruz Island, Galapagos. *Bot. J. Linn. Soc.* **78**, 67–84.

Hamann, O. (1984) Plants introduced into Galapagos—not by man, but by El Niño? *Noticias de Galapagos* No. 39, 15–19.

Hamann, O. (1985) The El Niño influence on the Galapagos vegetation. In: Robinson, G. and del Pino E.M. (Eds), *El Niño in the Galapagos Islands: The 1982–1983 Event*, Charles Darwin Foundation for the Galapagos Islands, Quito, Ecuador, pp. 299–330.

Harris, M. (1974) *A Field Guide to the Birds of Galapagos*, Collins, London.

Jordán, R. (1983) Preliminary report of the 1982–83 Niño effects in Ecuador and Peru. *Trop. Ocean-Atmos. Newsl.* No. 19, 8–9.

Kelly, R. (1985) Aspectos generales de El Niño 1982–83. *Invest. Pesq.* (special issue) **32**, 5–7.

Laurie, W.A. (1990) Effects of the 1982–83 El Niño sea warming on marine iguana (*Amblyrhynchus cristatus* Bell, 1825) populations on Galapagos. In: Glynn, P.W. (Ed.). *Global Ecological Consequences of the 1982–83 El Niño-Southern Oscillation*, Elsevier, Amsterdam.

Lawesson, J.E. (1988) The stand-level dieback and regeneration of forests in the Galápagos islands. *Vegetatio* (in press).

Leveque, R., Bowman, R.I. and Billeb, S.I. (1966) Migrants in the Galapagos area. *Condor* **68**, 81–101.

McMullen, C.K. (1986) Seed germination studies of selected Galapagos islands angiosperms. *Noticias de Galapagos* No. 44, 21–4.

Philander, S.G.H. (1983) El Niño Southern Oscillation phenomena. *Nature* **302**, 295–301.

Porter, D.M. (1983) Vascular plants of the Galapagos: origins and dispersal. In: Bowman, R.I., Berson, M. and Leviton, A.E. (Eds), *Patterns of Evolution in Galapagos Organisms*, Pacific Division, AAAS, San Francisco, CA, pp. 33–96.

Quinn, W.H., Neal, V.T. and Antunez de Mayolo, S.E. (1987) El Niño occurrences over the past four and a half centuries. *J. Geophys. Res.* **92**, 14449–61.

Rasmussen, E.M. and Wallace, J.M. (1983) Meteorological aspects of the El Niño/Southern Oscillation. *Science* **222**, 1195–202.

Robalino, M. (1985) Registros meteorológicos de la Estación Cientifica Charles Darwin para 1982–1983. In: Robinson, G. and del Pino, E.M. (Eds), *El Niño en las islas Galapagos. El Evento de 1982–1983*. Fundacion Charles Darwin para las Islas Galapagos, Quito, Ecuador, pp. 83–90.

Robinson, G. and del Pino, E.M. (Eds) (1985) *El Niño en las islas Galapagos. El Evento de 1982–1983*. Fundacion Charles Darwin para las Islas Galapagos, Quito, Ecuador.

Snell, H.L. and Christian, K.A. (1985) Energetics of Galapagos land iguanas—a comparison of 2 island populations. *Herpetologica* **41**, 437–42.

Snell, H.L. and Tracy, C.R. (1985) Behavioral and morphological adaptations by Galapagos land iguanas (*Conolophus subcristatus*) to water and energy requirements of eggs and neonates. *Am. Zool.* **25**, 1009–18.

Trillmich, F. and Limberger, D. (1985) Drastic effects of El Niño on Galapagos pinnipeds. *Oecologia* **67**, 19–22.

Wallace, J.M. (1985) Atmospheric response to equatorial sea surface temperature anomalies. In: Wooster, W.S. and Fluharty, D.L. (Eds), *El Niño North*. Washington Sea Grant Program, University of Washington, pp. 9–21.

Wiggins, I.L. and Porter, D.M. (1971) *Flora of the Galapagos Islands*. Stanford University Press, Stanford, CA.

Wooster, W.S. and Fluharty, D.L. (Eds) (1985) *El Niño North*. Washington Sea Grant Program, University of Washington.

Wyrtki, K. (1982) The Southern Oscillation, ocean–atmosphere interaction and El Niño. *Mar. Technol. Soc. J.* **16**, 3–10.

Wyrtki, K. (1985) Pacific-wide sea level fluctuations during the 1982–83 El Niño. In: Robinson, G. and del Pino, E.M. (Eds), *El Niño en las islas Galapagos. El Evento de 1982–1983*. Fundacion Charles Darwin para las Islas Galapagos, Quito, Ecuador, pp. 29–48.

2 Deforestation in the Tropics: Evaluation of Experiences in the Amazon Basin Focusing on Atmosphere–Forest Interactions

E. MEDINA
Centro de Ecología, IVIC, Aptdo 21827, Caracas 1020-A. Venezuela

2.1 INTRODUCTION

Deforestation is taking place at a fast rate throughout the tropics, even though there are substantial differences among continents and countries (Lanly, 1982). Large-scale transformation of tropical forests into agricultural fields and pastures (including the ongoing activities of slash and burn agriculture) are of great concern because of their potential impact on soil fertility and biological diversity. Effects on soil fertility lead to a reduction in the productive capacity of natural and artificial systems. Effects on biodiversity result in an impoverishment of the gene pool and a potential reduction of the capacity for restoration of forest composition and function. In addition, attention is being given to the relationship between deforestation of large tracts of tropical forests and the composition of the atmosphere. The main issues are:

1. Tropical deforestation is thought to contribute about 20% of the total carbon released into the atmosphere as a result of human activities in the biosphere (Detwiler and Hall, 1988).
2. The balances of nitrogen and sulphur compounds between the forest and the atmosphere are supposed to be significantly altered as a result of changes of patterns of land use in the tropics (Keller *et al.*, 1986; Andreae and Andreae, 1988; Wofsy *et al.*, 1988; Livingston *et al.*, 1988).
3. Elimination of large patches of forest, mainly in the Amazon basin, may have profound influence on the patterns of rainfall distribution and energy balance at local, continental, and even global scales (Salati *et al.*, 1978; Salati, 1987).

In this chapter I discuss why and how fast deforestation is taking place in the tropics; the actual changes in the nutrient balances of tropical forests induced

Ecosystem Experiments. Edited by H.A. Mooney *et al.*

by deforestation of different intensities; and the recovery capacity of tropical forest ecosystems after disturbance. I will explore the consequences for the global carbon balance of large-scale deforestation and the potential influence of secondary succession in reducing the carbon emissions caused by different types of forest conversion. Finally, I intend to highlight, based on examples drawn mostly from the Amazon basin, processes considered to be essential for the future design of ecosystem experiments.

2.2 DEFORESTATION IN THE TROPICS: WHY, WHERE, AND HOW FAST?

Underlying causes for high rates of deforestation in the tropics are clearly understood and practically all of them are related to economic development of third world countries (Fearnside, 1987). In many ways this process is similar to the 'age of large deforestations' which eliminated most natural forests in western Europe during the middle ages (Sternberg, 1973) or to the extensive deforestation of the eastern half of the United States which eliminated more than 99% of the natural forests by the end of World War II (Terborgh, 1975).

Population increase put more pressure on land for food and wood products, mostly with procedures of low technological level which are inappropriate to support dense populations, and are characterized by utilization periods of short duration, for instance increased utilization of slash and burn plots, and selective logging for local demands. At national levels economic need for development has led to the expansion of the agricultural frontier for the purpose of large-scale increases, both in cash crop and wood products for local consumption and export.

Alternatives to curb the rate of disappearance of tropical forests are necessarily associated to the solution of socioeconomic problems of tropical countries, therefore short-term practical solutions are not on the horizon yet. Ecologically minded attempts to tackle the deforestation problem in the tropics may be summarized as follows:

1. Development of more efficient agro-ecosystems for the whole range of climate variability in the tropics, capable of high outputs to cover the increasing requirements of growing populations.
2. Development of forest plantations capable of covering the increasing requirements for wood for local consumption and exportation.
3. Implementation of aggressive programs for the recovery of heavily disturbed and degraded ecosystems to be put into production or managed to accelerate secondary succession leading to formation of secondary forests.

These goals, however, are expensive to implement on a large scale and require well-orchestrated international cooperation, which has proven to be difficult to put into practice.

There are disagreements on the actual rate of tropical forest conversion (Fearnside, 1982; Lugo and Brown, 1982). Differences in estimated values of the areas involved derive from different definitions of forest conversion. For some authors, deforestation is the complete clearing of the forest, while selective logging is considered as utilization of productive forests without destruction (Lanly, 1982). For others, any intervention of tropical forests results in significant ecological changes, mainly because of its impact on biological diversity. Independent of these differences in definition of forest conversion, the rates of deforestation are substantial. Reliable measurements published by FAO (Lanly, 1982) amount to 1.13×10^6 km^2 year^{-1} for the whole tropical belt. In Latin America the rate of forest clearing accounts for about 36% of total deforestation in the tropics, but the rates are projected to increase in several countries (Table 2.1).

Table 2.1 Deforestation rates in the American tropics between 1976 and 1985 (from Lanly, 1982). Areas in km^2/year$\times 10^3$

Area	Rate 1976–80	Projected rate 1981–85
Mexico	5.3	5.95
Central America and Caribbean islands	4.0	4.26
Tropical South America	31.89	43.39
Total	41.19	43.39

For the Amazonia of Brazil Fearnside (1982) gives a rate of deforestation of about 16×10^3 km^2 year^{-1} which is converted to agricultural production. Malingreau and Tucker (1988) assessed the occurrence of large-scale deforestation (essentially elimination of forest canopy) in the southeastern Amazon basin using meteorological satellite data and ground observations. These authors showed that rates of forest disturbance and deforestation have increased markedly during 1984–85 and the rate of increase for the period 1975–85 appears to be exponential. In the States of Mato Grosso, Rondonia, and Acre total deforested area amounts to 89×10^3 km^2, while the disturbance area, where active deforestation is taking place, is approximately 265×10^3 km^2.

Rates of forest clearing vary considerably among and within countries, and in some of them there are active afforestation (or reforestation) programs which partially counteract the reduction of tree-covered areas caused by forest clearing (Table 2.2).

Table 2.2 Comparative clearing and afforestation rates in selected countries in the American tropics (from Lanly, 1982). Values in km^2/ year

Country	Deforestation rate	Afforestation rate
Cuba	20	110
Venezuela	1250	139
Mexico	5300	172
Brazil	14 800	3460

Table 2.3 Area of undisturbed and converted forests in the tropics ($km^2 \times 10^6$) (from Lanly, 1982)

	Logged	Fallow	Deforestation rate
Closed forest 9.85	2.1	2.39	0.075
Open forest 7.34	—	1.7	0.038
Total forests 17.19	2.1	4.09	0.113

It is important to recognize that all disturbed forest in the tropics are not completely destroyed by utilization, be it by logging or by transformation into agricultural land. A substantial area of disturbed forest is left as fallow (Table 2.3) and, as will be shown later, secondary succession accumulates organic matter at rates of several tons per hectare per year, particularly in humid, nutrient-rich sites, thereby counteracting the impact of carbon release caused by the initial perturbation.

2.3 ECOLOGICAL CONSEQUENCES OF DEFORESTATION IN THE TROPICS

Impacts of deforestation of inventories of living biomass, and reductions in soil fertility and biological diversity have been documented throughout the tropics (Sternberg, 1973). However, the degree of irreversibility of the human-induced changes is still debated. Experience suggests that evaluation of ecological consequences of deforestation has to take into account the intensity of the disturbance, climatic conditions, and the edaphological properties of the systems where intervention is taking place. Extrapolation to the whole tropical belt from results obtained in a given area in the tropics might be misleading due to climatic and geological variability which is expressed in the differences of biomass inventories and productive capacity documented for the tropics (Brown and Lugo, 1982).

2.3.1 INTENSITY OF DISTURBANCE AND RECOVERY MECHANISMS

Disturbance of tropical forests may be natural or human-induced, and may vary greatly in intensity. Natural disturbances of low impact are those caused by tree fall resulting from natural death or gusty winds which blow down areas of up to a few hectares. Severe disturbances may be caused by natural catastrophes such as hurricanes in the Caribbean, or landslides in tropical montane forests. Human-induced changes are also variable in intensity. Low disturbance intensity is related to slash and burn agriculture, by which areas of about 1 ha are slashed and burned after periods of dry weather. These areas are cultivated for three to five years and then abandoned, normally for periods longer than 20 years before reutilization. Higher demands for wood and cash crops as a result of increasing population densities have caused the elimination of larger forest chunks, frequently with heavy machinery, for development of extensive agricultural fields or for the establishment of permanent pastures for cattle raising. Intensity of disturbance differentially affects the recovery mechanisms of ecosystems. Disturbance includes physical and biological components. Physically it affects nutrient availability by increasing soil erosion and leaching. Biological impact includes partial destruction of the nutrient-conserving mechanisms operating in the forest, thereby reducing or eliminating altogether the re-establishment of the original vegetation in a secondary succession of variable duration. Uhl (1982) provided a systematic assessment of the relationship between intensity of forest disturbance and the recovery mechanism of plant communities (Table 2.4). Increasing disturbance intensity progressively eliminates forest recovery mechanisms, resulting in longer times for reforestation of the area.

Large-scale fires were thought to be of minor significance in humid tropical forests because of the lack of sufficiently long dry periods to render flammable accumulated dead biomass. However, the widespread practice of slash and burn agriculture in the humid tropics is a demonstration that burning is possible even after a few days of drought experienced in many of these areas. As seasonal drought occurrence increases so does the probability of fires. The recent catastrophic burning of thousands of hectares of undisturbed seasonal rain forest in southeast Asia showed how drastic fire can be during eventual episodic dry periods in the tropics (Leighton, 1984).

Prolonged dry periods during the Pleistocene have been shown for the Amazon basin (Absy, 1982; Dickinson and Virji, 1987). The widespread

Table 2.4 Common disturbance types in the lowland neotropics and the mechanisms of initial vegetation re-establishment characteristics of each (after Uhl, 1982)

Type of disturbance	Regeneration mechanisms				
	Advance regeneration	Sprouting	Seed Bank	Seed immigration from forest	Seed immigration from disturbed site
Single tree fall Forest trees	×	×	×	×	×
Multiple tree fall Forest trees		×	×	×	×
Forest cut and burn Successional trees and shrubs			×	×	×
Forest cut and burn followed by short term land use Grasses, forbs, succession trees, shrubs				×	×
Forest cut and burn followed by long-term land use Grasses					×

occurrence of charcoal at different depths in soils of the northern Amazon basin has been recently documented (Sanford *et al.*, 1985; Saldarriaga *et al.*, 1986). ^{14}C dating of charcoal remains give evidence of fires as old as 6×10^3 years in some tierra firme forest sites. One of the tierra firme forests investigated by Saldarriaga *et al.* (1986) has been burned twice during the past 600 years. These fires probably explain the patchiness in biomass and forest composition observed today. The distribution of radiocarbon dates of charcoal reported by Saldarriaga indicate events at 250, 1250, 2100, 2600, and 3100 years before present, and those peaks are probably associated with the occurrence of dry periods during the last 5×10^3 years in the Amazon (Saldarriaga *et al.*, 1986).

2.3.2 CHANGES IN BIOMASS AND NUTRIENT BALANCES AS A CONSEQUENCE OF SLASH AND BURN AGRICULTURE: SHORT-TERM IMPACTS AND RATE OF RECOVERY DURING SECONDARY SUCCESSION

Slashing a rainforest and burning the residues means an almost complete elimination of living above-ground biomass (40–160 Mg ha^{-1} for open and closed forests respectively—Brown and Lugo, 1982). All of this biomass is not reduced to CO_2 during the first year of perturbation. The stem biomass not burned and the underground roots are decomposed in periods lasting up to five to ten years. The immediate impact of slash and burn is the release of about 25% of the accumulated carbon in living and dead biomass (Palm *et al.*, 1986). The formation of ashes results in a sudden increase in the amount of available bases and P in the soil (Uhl and Jordan, 1984). This pulse of nutrients has been shown to be the base for crop and pasture production after forest clearing and burning (Falesi, 1976; Bushbacher, 1984; Uhl, 1987). However, this is a pulse of short duration, and after three to four years nutrient availability in slash and burn agriculture tends to return to pre-disturbance values (Uhl and Jordan, 1984).

After slashing, the soil becomes the most important repository for nutrient and organic matter in the ecosystem until secondary succession begins the process of new accumulation after farm abandonment. During cultivation the concentrations of total P and N in the system remain similar to that of the original forest, while Ca increases significantly during the first two to three years. Potassium is lost during longer periods and total losses are generally very large (Herrera *et al.*, 1981).

Recovery of biomass takes place at a rate ranging from 5 to 9×10^6 ha^{-1} $year^{-1}$ in nutrient-poor areas of the northern Amazon basin (Uhl, 1987). These rates do not differ significantly from those measured in secondary forests in richer soils in Central America (Ewel, 1971), a fact that makes necessary a reconsideration of the assumed nutrient limitation for secondary

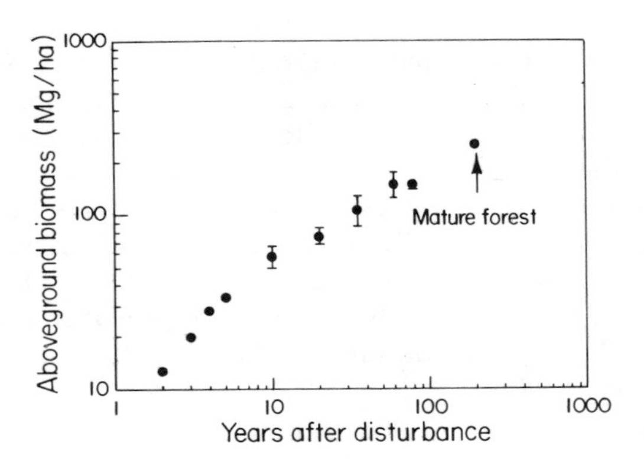

Figure 2.1 Composite of biomass accumulation data in secondary forests originated after slash and burn in the region around San Carlos de Río Negro in Colombia and Venezuela (data from Uhl and Jordan, 1984, and Saldarriaga *et al.*, 1986).

succession in the tropics (Harcombe, 1980). Net above-ground productivity reaches values similar to the undisturbed forest within five years of secondary succession (Uhl, 1987). Accumulation of biomass in farmed areas after prolonged periods of abandonment shows a nearly linear increase in total biomass between 10 and 80 years of fallow, (Fig. 2.1). It has been estimated that it takes nearly 200 years to reach biomass values similar to those of the original forest (Saldarriaga *et al.*, 1986).

Total ecosystem nutrient stocks recover at similar or faster rates than biomass during early succession (Uhl and Jordan, 1984). Within five years of secondary succession Ca and Mg are accumulated to around 70% of the original forest, K and N accumulated at a rate slightly faster than biomass, but P accumulated at a similar rate as biomass (Table 2.5).

It appears that slash and burn agriculture with fallow periods longer than 20 years, as practiced in areas of low population density, does not destroy the recovery mechanisms of forest ecosystems. A different situation arises when

Table 2.5 Percentage recovery after five years of secondary succession (Uhl and Jordan, 1984)

Biomass	36
Nitrogen	43
Phosphorus	33
Potassium	44
Calcium	67
Magnesium	75

dense human populations obtain their subsistence from this type of forest utilization, as in southeast Asia, where fallow periods of less than 20 years appear to be common (Palm *et al.*, 1986).

2.3.3 CHANGES IN BIOMASS AND NUTRIENT BALANCES AS A CONSEQUENCE OF FOREST CONVERSION TO PASTURE

Perhaps the most common land-use system determining clearing of large patches of Brazilian Amazon forests is the formation of pastures for cattle raising (Fearnside, 1987). Soil fertility after transformation to pasture in the central Amazon basin increased after forest clearing. Higher nutrient availability is maintained for periods of up to 11 years, provided that management of pastures avoids overgrazing (Falesi, 1976). Soil organic matter, pH, and available bases remains high for a number of years (Fig. 2.2). The availability of P increases markedly during a few years after conversion, but comes back to the original forest level within five to six years.

Analyses of the soil profiles from the same areas showed that the 'fertilization' effect resulting from deforestation was apparent mostly in the very top soil layers, particularly for exchangeable bases (Fig. 2.3). In the case of P the result of deforestation was a clear availability reduction in the soil layers below 20 cm.

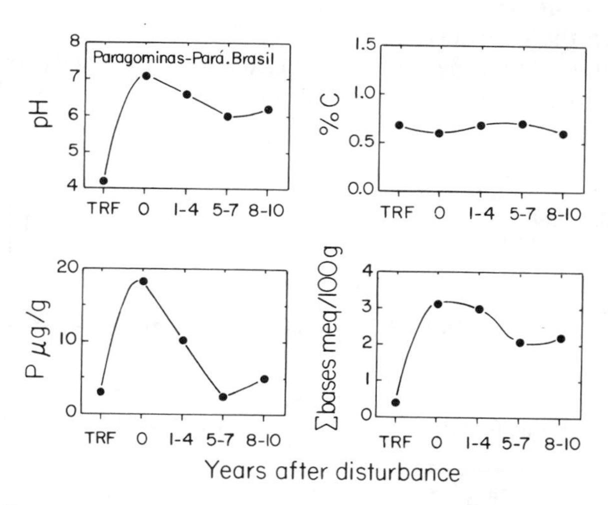

Figure 2.2 Changes in chemical properties of the upper 20 cm of soil after several years of forest conversion to pasture in the Amazon basin (TRF = tropical rain forest) (data from Falesi, 1976).

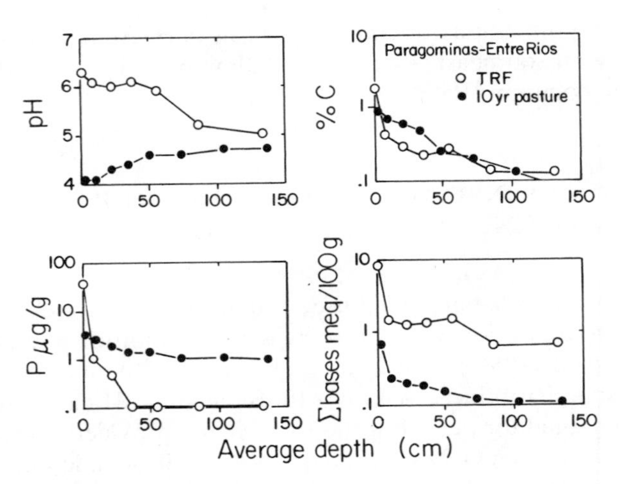

Figure 2.3 Profile of soil chemical properties of undisturbed rainforest and pasture areas after ten years of conversion (data from Falesi, 1976).

A detailed study of the nutrient dynamics after forest conversion to pasture in poor soils of the upper Rio Negro showed that maintenance of higher soil fertility than in forest soils is solely due to the continued release of nutrients from organic matter remaining above and below ground after conversion (Bushbacher, 1984). Even after two years of development pastures formed with *Brachiaria decumbens* lose significantly higher amounts of nutrients compared with the undisturbed forest (Fig. 2.4).

As a result of these net losses, and the lower capacity of grasslands compared to forests to accumulate large stocks of biomass, the nutrient

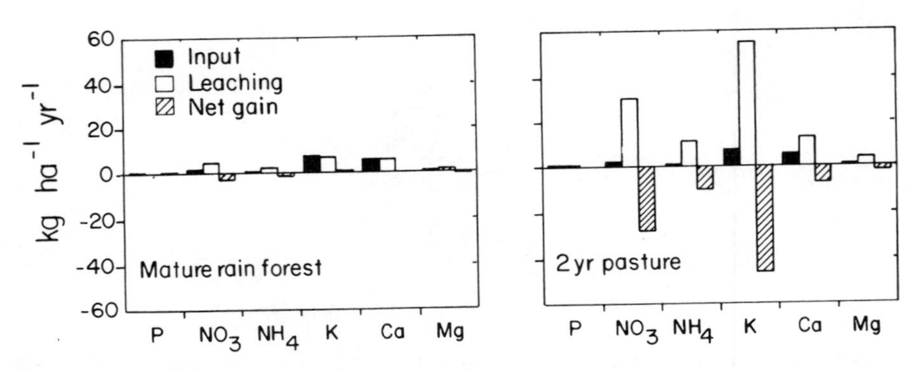

Figure 2.4 Nutrient balance in soils of undisturbed rainforest and pasture after two years of conversion (data from Bushbacher, 1984).

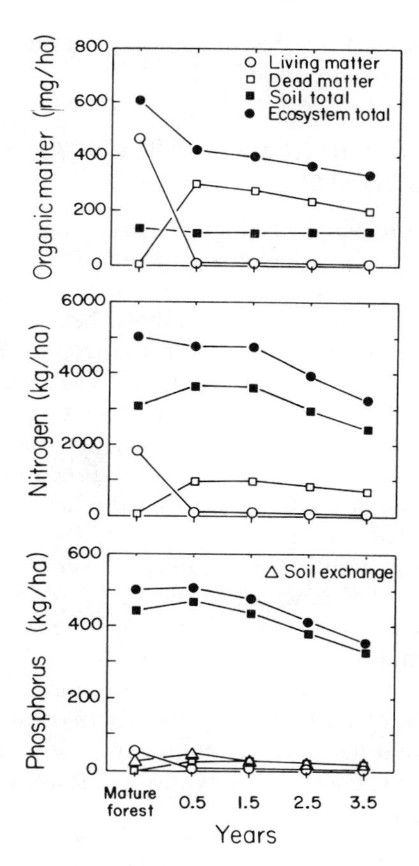

Figure 2.5 Changes in biomass, N and P stocks in a rainforest ecosystem in San Carlos de Río Negro after various periods of forest conversion (data from Bushbacher, 1984).

inventory of the ecosystem is reduced markedly in a matter of few years (Fig. 2.5).

As expected, the productivity of cultivated pastures in deforested areas is reduced strongly within 5–10 years and requires heavy additions of P to maintain reasonable rates of organic matter production (Serrão *et al.*, 1979). The maintenance of improved pastures on most Amazonian soils characterized by high P fixation capacity is heavily dependent on fertilization and conditioning of the soils (Fenster and León, 1979).

Conversion of forests to pasture, and maintaining pastures into production, seem to be a wasteful land-use system in regard to nutrient balance and gaseous exchange with the atmosphere. These pastures, which are not self-maintained, are probably net sources of carbon, and most probably also of nitrogen oxides, for prolonged periods.

2.3.4 EFFECTS OF DEFORESTATION ON LOCAL AND GLOBAL WATER BUDGETS

Impact of deforestation on local hydrological budgets has been shown in large watersheds. The main effect is the increase of runoff from the affected areas as a result of reduced evapotranspiration as well as soil compaction impeding deep drainage. Changes in rainfall patterns have been also postulated as a possible result of large-scale deforestation in the tropics. Demonstration of this possible effect is difficult. For the whole of the Amazon basin (around 7×10^6 km^2) Brazilian researchers have been able to show by measuring isotopic composition of rain water (^{18}O and D) that the oceanic humidity sources are progressively diluted in the east–west direction as a consequence of evapotranspiration (Salati *et al.*, 1978; Matsui *et al.*, 1983; Salati, 1987). For the whole Amazon basin rainfall is given as 12.0×10^{12} m^3 year^{-1}, while discharge measured at the mouth of the Amazon river has been estimated in 5.5×10^{12} m^3 year^{-1}. The difference of 6.5 m^3 year^{-1} corresponds to evapotranspiration. This amount has been confirmed by different methods. Several authors have shown that the continuous equatorial forest in Amazonia transpires about 50% of the total rainfall. Isotopic composition studies indicate that this amount of water is being recirculated within the basin itself (Salati, 1987). Therefore it has been concluded that large-scale deforestation not only will increase total runoff, but may result in lower total rainfall, and a more pronounced seasonality.

2.3.5 DEFORESTATION AND THE ATMOSPHERIC CARBON BUDGET

Recent models of the carbon budget of the atmosphere agree on an average net carbon release from clearing of tropical forests of 1×10^{15} g C year^{-1} in 1980, about 16% of the total human-induced carbon release to the atmosphere in that year (Table 2.6) (Kohlmaier *et al.*, 1985; Detwiler and Hall, 1988). Since undisturbed terrestrial ecosystems are considered neutral in regard to the carbon budget, there are 1.2×10^{15} g C year^{-1} which remained to be accounted for by these models. That is, there is a need for a large sink of carbon, probably located in terrestrial ecosystems, since ocean models do not yet allow for higher uptake rates. This large amount of carbon may be at least partially accounted for if the following processes are taken into consideration:

1. The extent of forest conversion has been overestimated. This uncertainty is being reduced with the use of satellite imagery coupled with ground

Table 2.6 Global budget of carbon for 1980 (Detwiler and Hall, 1988)

Sources and sinks	Flux ($\times 10^{15}$ g carbon per year)		
	Extreme	Median	Extreme
Released			
Fossil fuel combustion; cement production	4.8	5.3	5.8
Tropical forest clearing	0.4	1.0	1.6
Non-tropical forest clearing	−0.1	0.0	0.1
Accounted for			
Atmospheric increase	−2.9	−2.9	−2.9
Ocean uptake	−2.5	−2.2	−1.8
'Missing'	−0.03	1.2	2.8

confirmation (Malingreau and Tucker, 1988), and with more careful estimations of tropical forests biomass and finer analyses of the process of biomass oxidation and decomposition (Palm *et al.*, 1986; Detwiler and Hall, 1988).

2. Natural forests are not neutral, but constitute sinks for carbon, which accumulates in soil and wood or is exported through river systems to the ocean (Lugo and Brown, 1986). Amounts of carbon exported from terrestrial ecosystems can be considerable, and have been recently estimated to be in the range of $0.59–0.91\times10^{15}$ g C year^{-1} (Degens and Ittekkot, 1985).
3. The importance of secondary forests, forest plantations, and of pastures as carbon sinks has not been properly assessed.

From the results on rates of biomass accumulation in secondary succession in Amazonia forests of the upper Rio Negro (growing on very infertile soils) and in Central America (on relatively richer soils) an approximate figure for the amount of carbon being sequestered by secondary forests can be calculated. Above-ground biomass accumulating during the first 10 years of fallow ranges between 5 and 9×10^{8} g km^{-2} year^{-1} (Ewel, 1971; Uhl, 1987; see Fig. 2.3). Taking an average of these figures, and considering carbon content of organic matter as 0.50, it can be calculated that those secondary forests are sequestering 3.5×10^{8} g C km^{-2} year^{-1}. Dividing the amount of 'missing' carbon (1.2×10^{15} g C year^{-1}) by the rate of carbon fixation in above-ground biomass during early succession, gives an area of 3.4×10^{6} km^{2}. Lanly's (1982) figures indicate that by 1980 there were about 2.4×10^{6} km^{2} as forest fallow originated from logging or deforestation of closed forests. From open forests, which certainly have lower recovery rates during secondary succession, an additional amount of 1.7×10^{6} km^{2} was classified as forest fallow for the same year. If we add to those figures the rate of accumulation as underground biomass, we can see that the present area of disturbed vegetation, without further human intervention, can probably absorb the 'missing' carbon.

Plantation forests in the tropics can play also a significant role in compensating for the increases in atmospheric CO_2. There are about 0.11×10^6 km^2 of plantation in the tropics (Lanly, 1982), which show growth rates ranging from 1 to 30×10^6 g ha^{-1} year^{-1} (Lugo *et al.*, 1988), an amount that can be significant for the atmospheric CO_2 balance (Brown *et al.*, 1986). Plantation forests may also be instrumental in the recovery of degraded ecosystems, acting as sinks for carbon and nitrogen, and enriching the upper soil layers with nutrients scavenged from incoming rainfall or pumped from deeper soil layers.

Balancing that atmospheric carbon budget will continue to be a subject of debate for some time, until uncertainties in the estimation of the contribution of deforestation of tropical forests are eliminated, the role of secondary forests as carbon sinks is assessed precisely, and the influence of increased atmospheric CO_2 concentrations on net ecosystem production is definitively established. Recent information obtained from the analyses from air trapped in ice cores demonstrates that the simple approach of estimating areas of tropical forests being converted per year, and calculating from them the potential carbon release, will not contribute substantially to understanding the long-term process of carbon exchange between the atmosphere, the oceans, and the terrestrial ecosystems. Siegenthaler and Oeschger (1987) showed that the annual biospheric input between 1959 and 1983 ranged from 0 to 0.9×10^{15} g C year^{-1}, depending on the model used to describe the ocean–atmosphere interactions. Biospheric CO_2 releases were already high in the nineteenth century and no increase in the carbon release rate could be detected during the twentieth century.

2.4 DEFORESTATION AND BIODIVERSITY

While the contribution of tropical deforestation to atmospheric CO_2 can be at least partially offset by alternative carbon fixation processes in terrestrial ecosystems, or transportation and burial of organic residues in the sea, the question of biological diversity remains a serious issue which may be of more significance in the long range (Simberloff, 1986). It seems that estimations of loss of diversity in the tropics have been estimated as very large without a clear basis for its calculation (Lugo, 1988). However, reduction of the area of natural forests in the tropics represents a real threat for the survival of a large number of species. Tropical humid forests are without doubt the most diverse ecosystems in the world (Gentry and Dodson, 1987) with more than 50% of their plant species depending for survival on the presence of a continuous tree cover. Most tree species in the humid tropics are self-incompatible or dioecious (Bawa and O'Malley, 1987), depending on the maintenance of populations scattered over large areas for successful reproduction. Animal species are probably being forced into extinction because of the transformation of large forest areas in the tropics into islands where survival and reproduction probabilities are critically reduced (Terborgh, 1975). The subject of conservation of biodiversity in the tropics certainly requires more attention and appropriate resources for development of a conceptual basis which should help to devise efficient and practical management strategies in the near future.

2.5 ECOSYSTEM EXPERIMENTS AS A TOOL TO ASSESS THE IMPACT OF CLIMATIC CHANGE ON THE FOREST–ATMOSPHERE RELATIONSHIPS

This chapter discussed a set of relevant data to understand how ecological processes are being altered by the extensive elimination of the forest cover and the consequences for: (1) recovery capacity of forest ecosystems; and (2) changes in atmosphere–terrestrial ecosystem matter and energy exchange. It dealt mainly with processes of organic matter production, nutrient accumulation and leaching, and the cycles of water and CO_2 at the ecosystem level. Processes such as organic matter decomposition, biological nitrogen fixation, and those associated with the biological properties of the primary producers (e.g. lifespan, photosynthesis, respiration, nutrient uptake) were not considered in any detail. However, those processes have to be considered in the design of experiments at the ecosystem level to fully understand the impact of changes in atmospheric composition and rainfall chemistry on ecosystem structure and function, and to make predictions on their future development.

There are a few ecological studies at the ecosystem level in the tropics that could be considered truly ecosystem experiments. Two experiments which I know from personal experience are:

1. The El Verde project in Puerto Rico, which analysed the effect of high doses of gamma radiation on the destruction and recovery of a low montane forest (Odum and Pigeon, 1970).
2. The San Carlos project in Venezuela, which analysed the impact of primitive agricultural activities on the nutrient cycles and nutrient conserving mechanisms of lowland tropical rainforests (Herrera *et al.*, 1981, 1984).

In these experiments strong disturbances on a small scale (e.g. killing of the forest through gamma radiation or slash and burn) were inflicted on presumably pristine forests, and the changes in nutrient balance, rates of succession, and organic production were monitored over several years. Large data sets have been obtained on the mechanisms of secondary succession and the resistance of nutrient-conserving mechanisms to strong, sudden modifications of the structure of the system. Ecological models developed on these bases allow a predictive understanding of the impact of natural or man-made catastrophes on natural forest ecosystems (Odum, in Odum and Pigeon, 1970, pp. 1191–285; Herrera *et al.*, 1984).

Ecosystem experiments required by the SCOPE project have to be of a more subtle type. The environmental parameters of interest are increasing slowly but steadily, for instance average and extreme temperatures, atmospheric CO_2 concentration, or rainfall patterns. These experiments should incorporate, on a scale at the level of km^2, differential changes in environmental parameters that are slighly larger than the present rates of increase. The ideal and obvious, but impractical, procedure would be the construction of huge phytotrons enclosing large pieces of natural forest, in which CO_2 supply, rainfall composition, amount, and distribution, and temperature regime could be manipulated.

An alternative approach would be to focus on ecosystem processes sensitive to small environmental changes. Among those processes are rates of organic matter decomposition and soil respiration and the responses of epiphytic communities, richly represented in tropical forests (lichens and bromeliads seem to be particularly appropriate in the neotropics), and the ground flora. Experimental treatments may include modification of soil pH (to simulate processes of acidification or alkalinization), fertilization with limiting nutrients, changes in the chemical composition of rainfall (to investigate the effect of increased ionic load), and increases in the CO_2 gradients within the forest (to simulate the effect of increased organic matter decomposition). Effects on ecosystem function could be monitored using standard techniques, but the incorporation of new methodologies more appropriate for large-scale measurements is a must. Recent advances in techniques for measuring diatomic gases (for instance CO_2, CH_4, SO_2, NO_2) on a large scale have proved to be rather accurate and informative in the Amazon basin (Keller *et al.*, 1986; Livingston *et al.*, 1988; Wofsy *et al.*, 1988). In the same general area, stable isotopic techniques have been successfully applied to determine the origin of rainfall water (Matsui *et al.*, 1983) or to evaluate the recirculation of CO_2 from soil respiration (Medina *et al.*, 1986).

Presently the most difficult aspect of global change to examine through ecosystem experiments is the impact of changes in temperature. So far I know of no practical experimental design.

There are many feasible ecosystem experiments which, taking advantage of present development of ecological theory, could provide solid insight into the possible consequences of the present trends of global change. A few examples are listed in Table 2.7.

Table 2.7 Examples of simple ecosystem experiments to obtain a predictive understanding on the impact of present trends in atmospheric chemistry

Experimental treatments	Process to be monitored
Changes of composition and pH of incoming rainfall	Responses of epiphytic communities: composition, survival, water relations, photosynthesis, nutrient accumulation
Changes in frequency and intensity of rainfall	Analysis of canopy throughfall
Changes in gradients of CO_2 and other gases within the forest	Rates of soil respiration and organic matter decomposition
Fertilization with limiting nutrients	Photosynthesis of the ground flora and carbon isotopic composition of primary producers
	Gas exchange over the forest using micrometeorological techniques and airplanes

2.6 SUMMARY

Deforestation of tropical forests is taking place at a fast pace, conservative estimates are above 100×10^3 km^2 year^{-1}. There is evidence that, at least in some large forested areas of the world such as the southern Amazon basin in Brazil, rates of deforestation are increasing. Causes of deforestation are generally associated with development of economic schemes aimed to increase production of wood and cash crops, but also to provide a sustaining base for increasing populations in the countries involved. Alternatives to curb this trend are not in sight; the only hope resides in developing strong international cooperation, combining economy with ecology.

Ecological impacts of deforestation have been well documented throughout the tropics. Most are related to the loss of ecosystem recovery capacity as a consequence of the destruction of nutrient-conserving mechanisms operating in the native forest. The magnitude of the impact is associated with the intensity of the disturbance, original soil fertility and topographic characteristics of the areas affected. Several tropical forest ecosystems, in widely different soil types, have been shown to recover during periods varying between 80 and 200 years if the magnitude of the disturbance is low. Large-scale deforestation resulting from industrial logging operations or conversion to modern agriculture leads to profound changes, probably irreversible for all practical purposes. Conversion of tropical forests to pasture seems to be a wasteful land-use system in regards to nutrient balance and gaseous exchange with the atmosphere.

There is compelling evidence indicating that large-scale deforestation in the tropics is bound to affect the water cycling over extensive areas, and to decrease the capacity of the forest cover to regulate the CO_2 concentration of the atmosphere. The present controversy on the role of tropical forests in the global atmospheric carbon balance emphasizes the need for more accurate measurements of present rates of change in the forest cover, and the actual gas exchange over large pieces of continuous forests. Some of those measurements have been successfully performed in the Amazon basin using satellites and new techniques of gas exchange measurements using sensors mounted in airplanes. Another urgent need is the allocation of appropriate resources to evaluate the impact of deforestation on biodiversity.

Experiments at the ecosystem level have been conducted in several rainforests in the tropics. They have consisted of sudden and profound change in the structure of the forest. From these experiments a large database has been obtained leading to the development of testable models of the ecosystem structure and function. Ecosystem experiments to assess the effects of present trends in global change on the productive capacity and maintenance of tropical rainforests can be devised taking advantage of those theoretical developments. Such experiments include treatments proportionate to the present rates of change in atmospheric chemistry. Most aspects related to changes in atmospheric chemistry can be analyzed through relatively simple experiments. Testing temperature effects, however, presents a higher level of difficulty.

ACKNOWLEDGEMENTS

Acknowledgements are due to Drs Ariel Lugo (Institute of Tropical Forestry, USDA, Puerto Rico) and Victor García (Centro de Ecología, IVIC) for their comments and suggestions.

REFERENCES

Absy, M.L. (1982) Quaternary palinological studies in the Amazon basin. In: Prance, G.T. (Ed.), *Biological Diversification*, Columbia University Press, New York, pp. 67–73.

Andreæ, M.O. and Andreæ, T.W. (1988) The cycle of biogenic sulphur compounds over the Amazon basin. 1. Dry season. *J. Geophys. Res.* **93**(D2), 1487–97.

Bawa, K.S. and O'Malley, D.M. (1987) Estudios genéticos y de sistemas de cruzamiento en algunas especies arbóreas de bosques tropicales. *Rev. Biol. Trop.* (Costa Rica) **35** (Suppl. 1), 177–88.

Brown, S. and Lugo, A.E. (1982) The storage and production of organic matter in tropical forests and their role in the global carbon cycle. *Biotropica* **14**, 161–87.

Brown, S. and Lugo, A.E. (1984) Biomass of tropical forests, a new estimate based on forest volumes. *Science* **223**, 1290–3.

Bushbacher, R.J. (1984) Changes in productivity and nutrient cycling following conversion of Amazon rainforest to pasture, Ph.D thesis. University of Georgia, Athens, Georgia.

Degens, E.T. and Ittekkot, V. (1985) Particulate organic carbon—an overview. In: Degens, E.T., Kempe, S. and Herrera, R. (Eds), *Transport of Carbon and Minerals in Major World Rivers*, Part 3. SCOPE/UNEP Sonderband Heft 58, Mitt. Geol-Paläont. Inst. Univ. Hamburg, pp. 7–27.

Detwiler, R.P. and Hall, C.A.S. (1988) Tropical forests and the global carbon cycle. *Science* **239**, 42–7.

Dickinson, R.E. and Virji, H. (1987) Climate change in the humid tropics, especially Amazonia, over the last twenty thousand years. In: Dickinson, R.E. (Ed.), *The Geophysiology of Amazonia*, John Wiley & Sons, New York, pp. 91–101.

Ewel, J. (1971) Biomass changes in early tropical succession. *Turrialba* **21**, 110–12.

Falesi, I.C. (1976) *Ecossistema de pastagem cultivado na Amazonia brasileira*, Boletín Técnico No. 1, EMBRAPA. Belém.

Fearnside, P.M. (1982) Deforestation in the Brazilian Amazon: how fast is it occurring? *Interciencia* **7**, 82–8.

Fearnside, P.M. (1985) Brazil's Amazon forest and the global carbon problem. *Interciencia* **10**(4), 179–86.

Fearnside, P.M. (1987) Causes of deforestation in the Brazilian Amazon. In: Dickinson, R.E. (Ed.), *The Geophysiology of Amazonia*, John Wiley & Sons, New York, pp. 37–53.

Fenster, W.E. and León, L.A. (1979) Manejo de la fertilización con fósforo para el establecimiento y mantenimiento de pastos mejorados en suelos ácidos e infértiles de América tropical. In: Tergas, L.E. and Sánchez, P.A. (Eds), *Producción de pastos en suelos ácidos de los trópicos*, Centro Internacional de Agricultura Tropical, Cali, Colombia, pp. 119–34.

Gentry, A.H., Dodson, C. (1987) Contribution of nontrees to species richness of a tropical rain forest. *Biotropica* **19**, 149–56.

Harcombe, P.A. (1980) Soil nutrient loss as a factor in early tropical secondary succession. *Biotropica Secondary Succession*, pp. 8–15.

Herrera, R., Jordan, C.F., Medina, E. and Klinge, H. (1981) How human activities disturb the nutrient cycles of a tropical rainforest in Amazonia. *Ambio*, **10**, 109–14.

Herrera, R., Medina, E., Klinge, H., Jordan, C.F. and Uhl, C. (1984) Nutrient retention mechanisms in tropical forests: the Amazon Caatinga, San Carlos pilot project, Venezuela. In: DiCastri, F., Baker, F.W.G. and Hadley, M. (Eds), *Ecology in Practice, Part 1: Ecosystem Management*, Unesco, Paris, and Tycooy, Dublin, pp. 85–97.

Kaplan, W.A., Wofsy, S.C., Keller, M. and DaCosta, J.M. (1988) Emission of NO and deposition of O_3 in a tropical forest system. *J. Geophys. Res.* **93**(D2), 1389–95.

Keller, M., Kaplan, W.A. and Wofsy, S.C. (1986) Emissions of N_2O, CH_4 and CO_2 from tropical forest soils. *J. Geophys. Res.* **91**(D11), 11791–802.

Kohlmaier, G.H., Bröhl, H., Stock, P., Plöchl, M., Fischbach, U., Janecek, A. and Fricke, R. (1985) Biogenic CO_2 release and soil carbon erosion connected with changes in land use in the tropical forests of Africa, America and Asia. In: Degens, E.T., Kempe, S. and Herrera, R. (Eds), *Transport of Carbon and Minerals in Major World Rivers*, Part 3. SCOPE/UNEP Sonderband Heft 58, Mitt. Geol-Paläont, Univ. Hamburg, pp. 123–36.

Lanly, J. (1982) *Tropical Forest Resources*, Food and Agriculture Organization, Rome, FAO For. Pap. No. 30.

Leighton, M. (1984) Effects of drought and fire on primary rain forest in eastern Borneo. Abstract, AAAS Annual Meeting. AAAS Publication 84-4, Washington, DC.

Livingston, G.P., Vitousek, P.M. and Matson, P.A. (1988) Nitrous oxide flux and nitrogen transformations across a landscape gradient in Amazonia. *J. Geophys. Res.* **93**(D2), 1593–9.

Lugo, A.E. (1988) Estimating reductions in the diversity of tropical forest species. In: Wilson, E.O. and Peter, F.M. (Eds), *Biodiversity*, National Academic Press, Washington, DC, pp. 58–70.

Lugo, A.E. and Brown, S. (1982) Conversion of tropical moist forests: a critique. *Interciencia* **7**, 89–93.

Lugo, A.E. and Brown, S. (1986) Steady state terrestrial ecosystems and the global carbon cycle. *Vegetatio* **68**, 83–90.

Lugo, A.E., Brown, S. and Chapman, J. (1988) An analytical review of production rates and stem wood biomass of tropical forest plantations. *Forest Ecol. Man.* **23**, 179–200.

Malingreau, J.P. and Tucker, C.J. (1988) Large-scale deforestation in the Southeastern Amazon basin of Brazil. *Ambio* **17**, 49–55.

Matsui, E., Salati, E., Ribeiro, M.N.G., Reis, C.M. dos, Tancredi, A.C.S.N. and Gat, J.R. (1983) Precipitation in the Central Amazon basin: the isotopic composition of rain and atmospheric moisture at Belem and Manaos. *Acta Amazonica* **13**, 307–69.

Medina, E., Montes, G., Cuevas, E. and Roksandic, Z. (1986) Profiles of CO_2 concentration and $\delta^{13}C$ values in tropical rain forests of the upper Rio Negro basin, Venezuela. *J. Trop. Ecol.* **2**, 207–17.

Odum, H.T. and Pigeon, R.F. (1970) *A Tropical Rain Forest: a study of irradiation and ecology at El Verde, Puerto Rico*. Division of Technical Information, US Atomic Energy Commission, Oak Ridge, Tennessee.

Palm, C.A., Houghton, R.A. and Melillo, J.M. (1986) Atmospheric carbon dioxide from deforestation in Southeast Asia. *Biotropica* **18**, 177–88.

Salati, E. (1987) The forest and the hydrological cycle. In: Dickinson, R.E. (Ed.), *Geophysiology of Amazonia*, John Wiley & Sons, New York, pp. 273–96.

Salati, E., Marques, J. and Molion, L.C.B. (1978) Origem e distribuição das chuvas na Amazonia. *Interciencia* **3**, 200–6.

Saldarriaga, J.G., West, D.C. and Thrap, M.L. (1986) *Forest Succession in the Upper Río Negro of Colombia and Venezuela.* Oak Ridge National Laboratory, Environmental Sciences Division, Publication No. 2694.

Sanford, Jr R., Saldarriaga, J., Clark, K., Uhl, C. and Herrera, R. (1985) Amazonian rain-forest fires. *Science* **227**, 53–5.

Serrão, E.A.S., Falesi, I.C., Da Veiga, J.N. and Neto, J.F.T. (1979) Productividad de praderas cultivadas en suelos de baja fertilidad de la Amazonia del Brasil. In: Tergas, L.E. and Sánchez, P.A. (Eds), *Producción de pastos en suelos ácidos de los trópicos,* Centro Internacional de Agricultura Tropical, Cali, Colombia, pp. 211–43.

Siegenthaler, U. and Oeschger, H. (1987) Biospheric CO_2 emissions during the past 200 years reconstructed by deconvolution of ice core data. *Tellus* **39B**, 140–54.

Simberloff, D. (1986) Are we on the verge of a mass extinction in tropical rain forests? In: Elliott, D.K. (Ed.), *Dynamics of Extinction*, John Wiley & Sons, New York, pp. 165–80.

Sternberg, H.O. (1973) Desarrollo y conservación. *Bol. Soc. Ven. Cienc. Nat.* **22**, 427–59.

Terborgh, J. (1975) Faunal equilibria and the design of wildlife preserves. In: Golley, F.B. and Medina, E. (Eds), *Tropical Ecological Systems*, Ecological Studies No. 11, Springer Verlag, New York, pp. 369–90.

Uhl, C. (1982) Recovery following disturbances of different intensities in the Amazon rain forest of Venezuela. *Interciencia* **7**, 19–24.

Uhl, C. (1987) Factors controlling succession following slash and burn agriculture in Amazonia. *J. Ecol.* **75**, 377–407.

Uhl, C. and Jordan, C.F. (1984) Succession and nutrient dynamics following forest cutting and burning in Amazonia. *Ecology* **65**, 1476–90.

Wofsy, S.C., Harris, R.C. and Kaplan, W.A. (1988) Carbon dioxide in the atmosphere over the Amazon basin. *J. Geophys. Res.* **93**(D2), 1377–87.

3 What Ecological Lessons Can We Learn from Deforestation Processes in the Past?

C.O. TAMM
Department of Ecology and Environmental Research, Swedish University of Agricultural Sciences, P.O. Box 7072, S-75007, Uppsala, Sweden

3.1 INTRODUCTION

Conventional maps of world vegetation usually illustrate potential vegetation and thus give the impression that much of central Africa and southeast Asia are covered with rainforests. In the same way much of Europe, China, and North America east of Mississippi is depicted as belonging to the region of temperate deciduous forest.

We know that this is no longer true. Much of the original forest has been replaced by agroecosystems and strongly transformed woodlands. In many regions the forests still occurring are more likely to be plantations of exotic conifers or eucalypts rather than natural forests. Some of the once-forested land has become wasteland of little use for humans, such as the Imperata grasslands in the humid tropics or acidified or heavy-metal contaminated land around mines and industries.

Modern technique, in particular the use of satellites, is now able to offer a true picture of the world forest situation. Yet the old maps are still widely used. They are more beautiful, and even in the cases where the users realize that they show *potential* vegetation, both the users and the map authors often make the tacit assumption that present deviations from potential vegetation are reversible in all important respects.

If we are to make the global ecosystem a sustainable world we have to realize that worldwide vegetational changes are now occurring. We must determine their rates as well as their proximate and ultimate causes. Then we need to elucidate the environmental consequences of the changes. As deforestation and other vegetational changes are as old as agricultural man, we should ask the question whether we can learn from history. The purpose of this chapter is to discuss the potentials as well as the limititations of the historic approach.

Ecosystem Experiments. Edited by H.A. Mooney *et al.*

3.2 DEFORESTATION AND OTHER VEGETATIONAL TRANSFORMATIONS IN THE PAST AND PRESENT

Much early agriculture was based upon some system of shifting cultivation, which involved clearing of pre-existing vegetation and consumption of nutrient stores accumulated on the site by this vegetation. The impact on nature increased rapidly with the development of permanent settlements, as in the early cultures of the Middle East and Egypt. Unlike shifting cultivation the river valley agriculture based on irrigation was not dependent on the plant nutrients accumulated in forest ecosystems for food production, since flooding water and airborne dust replaced some of the mineral nutrients removed by harvest. Still forest products were needed as fuel, building material, fencing, etc. There must have been an early strong pressure on the forest resources in the eastern Mediterranean region, as well as in other countries with intensive agriculture in dry regions. Solomon needed cedar logs transported all the way from Lebanon for the temple in Jerusalem.

Agriculture on non-irrigated land needs replenishment of plant nutrients removed with harvests. Shifting cultivation was the earliest solution of the problem, still practised in some areas, but the system breaks down when increasing population pressure makes the fallow periods too short. Such overuse of land by shifting cultivation is now restricted to developing countries, but has certainly occurred widely, also in Europe, but not permanently, as it is a case of non-sustainable land use.

Yet agriculture on permanent fields also depended heavily on the forest land for replenishment of removed plant nutrients in many parts of the world. In Europe and around the Mediterranean Sea the most common method to transfer fertility from forest land to the cultivated fields was by heavy grazing, where the cattle, sheep, and goats left more of their droppings where they were milked or kept overnight than were they grazed. In areas with cold winters, in mountains and at higher latitude the supply of winter fodder became the bottleneck for agricultural land use, determining the number of cattle and sheep that could be supported. Much of the winter fodder was collected from relatively open woodlands by haymaking and pollarding. In addition forest litter was raked and used in cowsheds and stables for mixing with the manure—another transfer of plant nutrients from forest land to arable fields, common in middle Europe.

Various practices developed to increase the amount of fodder available from the forest land to domestic animals. Wooded pastures, mostly with deciduous trees in an open spacing, have been common in different parts of Europe. Often the trees were periodically pollarded, with the leaves used for fodder, directly or after drying. In not too densely populated areas there was often a combination of shifting cultivation in the forest and permanent

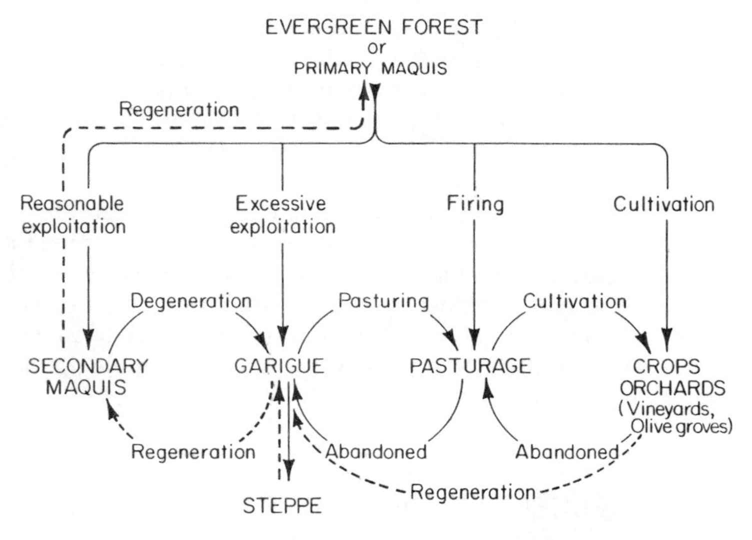

Figure 3.1 Stages in the degeneration and regeneration of plant communities, and the effect of man on Mediterranean vegetation. From Polunin and Huxley (1970). Reproduced with permission from Chatto & Windus Ltd, London.

agriculture on 'home' fields. When a burn-beaten area no longer supported a regular crop it might still provide grazing, and it also occurred that forests were burned to improve grazing. Even in cases where the burning was not intentional, the temporary improvement of the grazing was much appreciated. Many of the Mediterranean countries are now covered with various successional stages after fire (Fig. 3.1, from Polunin and Huxley, 1970).

In humid western Europe heavy grazing combined with burning led to the formation of open heathlands, areas where sheep and cattle could find something edible (grasses in summer and young *Calluna*, rejuvenated by periodic burning, all the year round).

Fire, whether natural or started by humans, certainly also played a great role in determining the kind of vegetation that dominated the transitional land between subhumid and subarid regions on all continents. In addition to fire, the number and species of large herbivores influence the forest–steppe boundary. Here humans have had a further indirect influence, first by hunting these animals, later by herding them and reducing their natural enemies.

The changes in agricultural systems in Europe during the past century, and particularly since World War II, have reduced the need to produce food and fodder from the forest land. The most decisive factor has been the access to chemical fertilizers, enabling the farmer to produce more food and fodder on the home fields with less labour. Another important factor has been the transformation of large areas of both woodlands and steppes on rich soils in North America and Australia into grain-exporting regions. A problem characteristic for the densely populated regions of the Old World, shrinking forest resources, was 'transferred' to other regions, with varying abilities to handle the environmental consequences.

For central and southern Europe, however, there have been many cases reported of increases in forest area and standing stock of trees during the past few decades, in contrast to what is now happening in many developing countries. The afforestation is both spontaneous and artificial (in the latter case often plantations of exotic species), and the soil has often been degraded to an extent making it unlikely that the new forest will resemble the old ones in productivity, or species composition.

3.3 FOREST INFLUENCES—BRIEF OVERVIEW

3.3.1 RADIATION BALANCE AND TEMPERATURE CLIMATE

It is a trivial remark to say that the microclimate is different in a forest, compared with in an open field. Yet standard meteorological data refer to conditions in the open field and not to conditions in a forest. While the daily amplitude in temperature is usually narrower inside the forest than outside it, the difference in mean temperature may not be very large. Frost frequency is usually lower under forest canopy, and daily maximum temperatures are usually higher in the open.

In cold climates, with much of the precipitation as snow, easily intercepted in tree crowns, different vegetation cover may cause ecologically important differences in soil temperature, even leading to permafrost beneath certain tree species (Drury, 1956; Viereck, 1970).

Forests usually have a lower albedo—absorb more and reflect less in the visible part of the incoming solar radiation—than most other vegetation types. As much of the energy comes with radiation of other frequencies, the albedo in the visible range may be of limited importance, but it is clear that snow-covered ground reflects more energy than snow-free ground. Where the presence of trees changes the dates for snow-melting, this certainly affects the local climate.

The direct effects of forests on the temperature climate consist of impacts on reasonably well-known physical processes, which can be studied on existing ecosystems. The historical approach is not a main avenue for this research.

3.3.2 EVAPOTRANSPIRATION, RUNOFF, SOIL PHYSICAL PROCESSES

Numerous catchment studies all over the world indicate that the presence of forests normally decreases runoff (Fig. 3.2, from Nemec and Rodier, 1979). At the same time runoff peaks are reduced and dry-season runoff may be increased. The causal background for the differences is to be found in higher evapotranspiration from forests because of their large transpiring leaf surface (which also intercepts much rain, rapidly evaporating after a shower) and longer transpiration period (particularly evident when evergreen forests are compared with arable land). Forests also have more permeable soils than arable land, and many types of grasslands, also often have deeper roots using a larger soil volume for their water supply.

The actual precipitation in a forest may be higher than in the open, as the rough surface of the forest canopy may capture fog and rain droplets which have a low sedimentation rate and therefore are measured very incompletely in conventional rain gauges. The effect is particularly evident in some mountain forests where 'fog-drip' may form a considerable part of the total

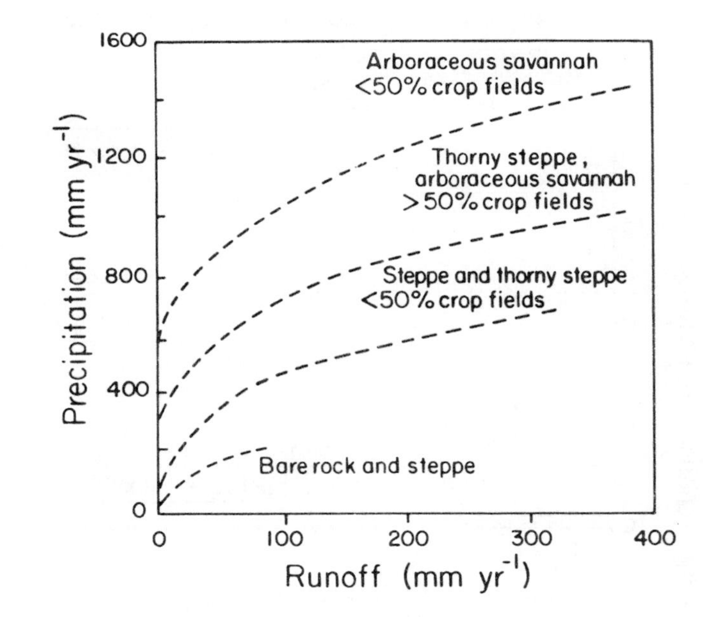

Figure 3.2 The relationship between annual precipitation and runoff as a function of vegetation. The relationship was developed on the basis of data collected by ORSTOM by Nemec and Rodier (1979). The range of each curve reflects the distribution of the available data. There were, for example, no data for runoff from bare rock at high precipitation.

precipitation. Also, forest edges may get more precipitation than both open fields and the interior of the forest. This more or less 'occult' precipitation can seldom change the general picture, viz. that afforestation decreases and deforestation increases runoff. Yet it should be kept in mind that precipitation measured in standard gauges does not reflect the heterogeneity in actual amount of water supplied to the ground, and that this heterogeneity is much larger in a forest (stem-flow, drip from branch terminal shoots in some tree species, etc.), and that the deviations from standard meteorological data are considerable in forest edges and generally in mosaic landscapes with both tree groves and grassy patches.

The high permeability of forest soils for water is connected with the high heterogeneity of both vegetation and soil. Vertical channels formed by large roots may persist long after the death of the root and the tree. Both forests and grasslands may also have channels formed by burrowing animals (earthworms, moles, and others), but grasslands are more subjected to soil compaction, e.g. by trampling by large herbivores. There is thus less surface runoff from a forest than from open fields. The transport of water through the unsaturated zone, and further with the groundwater gradients, occurs both in forest soils and outside the forest, but this process is much slower than surface runoff. The low proportion of surface runoff and high proportion of ground-

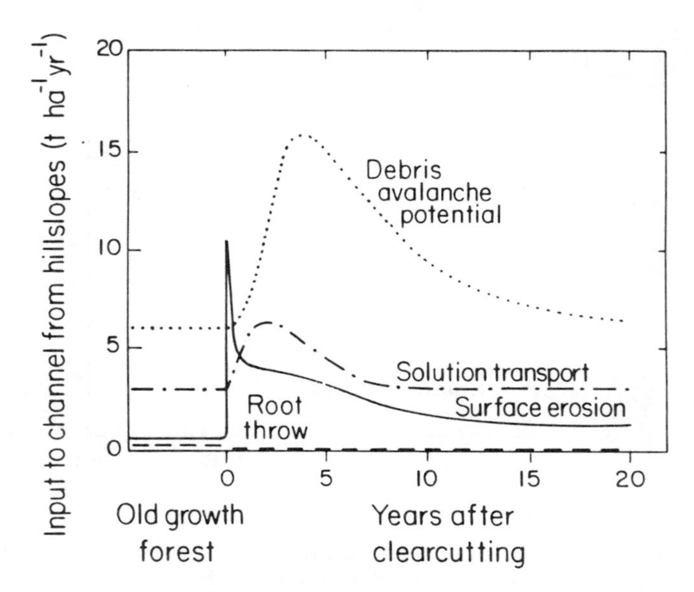

Figure 3.3 Changing probabilities of various transport mechanisms following logging of an old growth Douglas fir forest in Oregon, Pacific Northwest, United States. From Swanson *et al.* (1982).

water runoff in forests is the main reason why erosion is much lower in forested areas than in areas with high proportions of pastures and, in particular, arable land. A further reason why erosion is less from permanently vegetated areas is the protective effect of roots, both living and recently dead ones (Fig. 3.3, from Swanson *et al.*, 1982). The living vegetation and the forest litter layer also offer mechanical protection of the soil, when heavy raindrops first hit the vegetation and litter and thereby lose much of their kinetic energy.

As will be discussed later, changes in river transport of material are among the processes in the past which we can measure and relate to human impacts on the ecosystems.

3.3.3 ACCUMULATION AND DISSIPATION OF ORGANIC MATTER AND NUTRIENTS IN FOREST ECOSYSTEMS

Compared with most other ecosystems, forests accumulate large amounts of organic matter and plant nutrients. Part of these stores is bound in living biomass (in which the dead heartwood of living trees is also included, somewhat illogically). Another part is bound in the soil. While the total amounts of carbon on the site increase from cold taiga forests to the tropical rainforest, the proportion of the carbon accumulated in the soil decreases in the same direction. With decreasing humidity, the accumulation of carbon also decreases, reflecting the decreasing primary production.

It is characteristic for most forests that there is a change between accumulation phases and dissipation phases. In boreal forests and other forests with more or less even-aged tree stands, the accumulation dominates over most of the rotation time, while clearfelling or natural causes (fire, wind, insect outbreaks) kill the trees and start a period with intensive decomposition and nutrient mineralization. When new vegetation is established, accumulation starts again. The same processes are also active in other forest types, but in tropical rainforest and some types of deciduous hardwood forest there is often a more patchy distribution of sites where either accumulation or dissipation dominates. Most man-made forests are even-aged, and have the pattern of alternating phases built in. Despite the local changes between aggrading and degrading phases in a forest region, there is in the long run no net loss or gain of oxygen or carbon to the atmosphere, as long as the forested area remains stable (assuming that the carbon in the harvested forest products eventually is returned to the atmosphere by decomposition or burning). This is in contrast to what is often written in popular articles. On the other hand, changes from forests to ecosystem types with lower accumulation of organic matter mean an addition of carbon dioxide to the atmosphere. Even if the combustion of fossil fuels is considered to give the largest contribution to the observed increase in atmospheric carbon dioxide concentration, changes in land use, mainly deforestation and increased amount of arable land, can by no means be neglected (Woodwell *et al.*, 1983).

Nutrient stores are often more or less in proportion to the stores of carbon. This is particularly true for nitrogen on sites where the growth of the trees is nitrogen-limited. Primary production is in these cases proportional to available nitrogen, and even if there is a retention of nitrogen during litter decomposition, most of the nitrogen stays in organic form, bound to carbon. Other essential elements are also accumulated roughly in proportion to biomass, even if the geochemical behaviour of the various elements may cause deviations from proportionality to organic matter in the soil, e.g. in the case of phosphorus in soils rich in iron and aluminium oxides.

As stated earlier, the possibility of exploiting the accumulated stores of nutrients in forest ecosystems has been of great importance for agricultural man through millennia, so any possibility of tracing the consequences of this process should be welcomed.

3.3.4 SOIL CHEMICAL AND BIOLOGICAL CHANGES

Jenny (1941) defined five groups of 'soil-forming factors': parent material, topography, climate, organisms, and time, where time plays a role somewhat different from the other four groups that primarily determine the direction of soil profile development, while time determines the rate at which the state of a mature profile is attained.

Until the modern phenomenon of acid deposition most of the human interference with soil-forming processes has been achieved by changing the vegetation. As said earlier, these changes very often meant deforestation. Yet all soil changes take time, and it is therefore often possible to find traces of earlier processes in soil profiles by studies of mineral composition or morphological traits. Weathering pattern, weathering products, lateritization, occurrence of hardpans are some of the characteristics to look for, when we are trying to trace environmental effects of human impacts in the past.

3.3.5 BIOLOGICAL DIVERSITY

Most natural forests have much more biological diversity than agroecosystems and other ecosystems strongly transformed by humans. Yet many of man's earlier actions have increased diversity by creating a landscape containing a mixture of woodlands, grasslands, and arable fields. This landscape diversity has existed long enough for a biological evolution of ecotypes and in some cases species, adapted to the man-made environment. In addition, humans have distributed plant and animal species across natural barriers, with well-known effects on both productivity and diversity. There may be such changes related to the deforestation problems, but most effects of biological invasions concern primarily other topics, and are dealt with by SCOPE in previous syntheses, so they will be left out here.

3.3.6 SITE SUSCEPTIBILITY TO FIRE, EROSION, ETC.

Many anthropogenic land transformations result in ecosystems more sensitive than the original forest towards adverse factors such as fire and erosion. Heathlands, originally created by burning, are more fire-prone than the original hardwood forest. The same holds true for the Mediterranean scrub vegetation, maquis and garigue, compared with the original evergreen forest, even if the fire danger is high in all types of continuous vegetation in climates with long drought periods. In many parts of the world human action displaces the boundary between forests and fire-prone grasslands or savannahs in the direction of increasing area of the latter types.

Man-made forests, often coniferous, are on average more fire-prone than natural forests. This is not accidental, as a number of favoured conifer species (e.g. many *Pinus* species), are fire-tolerant pioneer species, either due to having thick and insulating bark, or to having reproductive mechanisms adapted to fire (serotinous cones, 'grass stages' of some pine species). Both crop plants and most plantation trees have traits of pioneer plants, with rapid early growth, often coupled with a demand for the nutritional improvements caused by site disturbance.

Fire, particularly when often repeated, opens the way for other types of site deterioration, nutrient depletion and erosion. The hardpan formation in much of the heathlands of the British Isles may not lead directly to erosion, but it does impede drainage and favours the growth of blanket bogs. There is considerable evidence that this development is caused not only by climate changes but also by human intervention.

3.4 ARCHIVES FOR ECOLOGICAL INFORMATION ON EFFECTS OF VEGETATION CHANGES

There are many sources of information on deforestation and other anthropogenic land transformations. Historical literature gives some information, e.g. that extensive forests covered much of present Germany at the time of Caesar and Augustus. The number of church towers visible from a hill on the island of Gotland when Linneaus travelled there in the 1750s shows that the landscape was far more open at that time than at present. Old paintings from Italy or The Netherlands often give the impression of a sparsely wooded and heavily grazed environment to the town or castle in the centre of the picture, independently of whether the motive came from ancient myths or from the artist's own country and time.

More detailed and also more reliable information can often be obtained from written sources such as tax accounts, royal charters, and legal acts when property was sold or divided up. The Domesday Book, dating from the Norman conquest of England in the eleventh century, is the outstanding example, but there are, in different countries, many types of archival documents containing ecological information not yet fully exploited.

From prehistoric time there is by definition no written information. However, archaeological excavations sometimes supply relevant ecological information: what sorts of domestic animals were present, what sources of food were used, and health conditions within the human population.

There are also biological and geological archives, some of which have long been used in vegetation history studies. Pollen analysis is now a classic method, and the information obtained tells us much about both changes in forest cover and the types of ecosystems which have replaced the original forest, such as heathland or arable land that are characterized by a number of easily recognizable weed pollen. The bias caused by the local environment to the peat bogs originally studied can be compensated by using lake sediments with less pollen from the immediate vicinity or by making the studies regional. Charcoal fragments in soils may reveal the species composition of the burned forest. Modern studies of lake sediments may also tell the acidity of the lake at different points in time (from diatom shells) and the frequency of forest fires in the catchment (from charred wood fragments). The dating can be checked using isotope methods.

In lake and sea sediments the rate of sedimentation *per se* is a source of information, as it is related to the rate of erosion on the surrounding land. A sediment core from the Black Sea shows low and uniform sedimentation rate from 800 BC until after the first centuries after Christ, when the general rate rises and the shifts between 'highs' and 'lows' become dramatic (Fig. 3.4, from Degens *et al.*, 1976). As the sediments of the Black Sea originate from many different types of ecosystems, it is not easy to pinpoint the role of deforestation for the pattern observed, but that the sedimentation reflects an increasing human influence appears very clear.

Annual rings in tree trunks are also an important source of information. The pioneer in this field, A.E. Douglass (1935; cf. however, Fritts *et al.*, 1965) concluded that a large prehistoric settlement in Mesa Verde, Colorado, was abandoned because of a prolonged drought AD 1273–85. It is tempting to draw parallels with the present situation in Africa, where there are strong reasons to believe that the drought, or at least much of the depletion of vegetation resources, is man-made. In the Mesa Verde example the rains eventually returned, to the advantage of the trees from which we have obtained the climatic record.

Annual layers occur not only in trees but also in other living organisms (e.g.

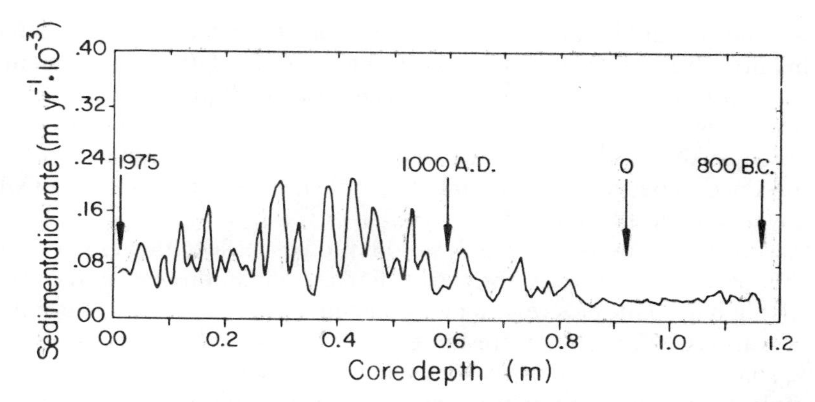

Figure 3.4 Rates of sedimentation in a sediment core from the Black Sea at a water depth of 470 m. The age assignment is based on varve counts and stratigraphic cross correlation with seven other Black Sea cores. From Degens *et al.* (1976).

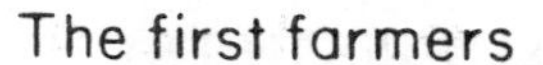

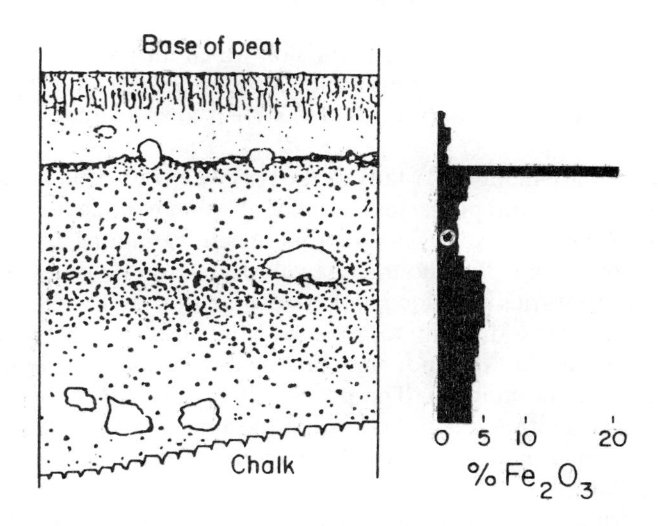

Figure 3.5 Double soil profile below peat at Goodland, Co. Antrim, Ireland. The graph on the right shows modest accumulation of iron in the enriched B horizon of the primary forest soil, and also the much greater quantity of iron in the 'iron-pan' B horizon of the podzol soil that formed later. From Proudfoot (1958).

mussel shells, Carell *et al.*, 1987), and in sediments (varved clays, certain lake sediments, glacier ice). Apart from offering accurate dating, the existence of annual layers shows that the sediment is undisturbed, and that movement of particles must have been minimal since the formation of the layer. Combined microscopic observations and use of modern chemical analytical technique can produce a considerable amount of information on the environment where the layer was once formed.

As shown earlier (section 3.3.4) soil profiles reflect the environmental conditions when they were formed. Chemical conditions (as well as pollen counts) in buried profiles (e.g. under grave-mounds) can provide information on conditions before they were buried. Because soil chemical processes are slow, some traits of earlier conditions may be preserved for long periods; for example accumulation of hydroxides of iron and aluminium deep in a former cambisol, while later accumulation has been concentrated in a thin hardpan nearer the surface (Fig. 3.5, from Proudfoot, 1958).

3.5 IS IT POSSIBLE TO COMBINE OUR HISTORICAL EVIDENCE WITH OUR PRESENT KNOWLEDGE OF FOREST INFLUENCES TO GET A RELIABLE PICTURE OF MAN'S ROLE FOR CHANGES IN THE ENVIRONMENT?

This is the central question for the discussion which this contribution is meant to start. I wish that I could give a simple yes as an answer, but that is not the case. Nor should we say that 'no' is the correct answer under all circumstances.

The problems are manifold. Many of the processes recorded in our archives are influenced by natural processes as well as by human impacts. The past few decades of research have revealed pronounced variations in temperature climate in historic time. The flourishing culture in Europe during the twelfth and thirteenth centuries coincided with a warm period, followed by the 'Little Ice Age', particularly difficult to cope with in already cold areas such as Scotland, Scandinavia, Iceland, and Greenland (where the Nordic settlements eventually succumbed). The previously mentioned prehistoric Indian settlement gives another example of the importance of climatic fluctuations.

Yet there is other evidence—in sediments, in soil profiles, and elsewhere—showing that humans have also acted as a geological factor in the past, e.g. by overuse of former forest land for pasturing and agriculture, thereby increasing sedimentation rate. The formation of the Atlantic heathlands, with its feedbacks on soil productivity and at least local climate, means an environmental change where man has played a key role. The timberline in densely populated mountains, such as the Alps and the Pyrenees, is to a large extent depressed by grazing, particularly on south-facing slopes where the insolation is higher (and where the prospects of adequate grazing after clearing are good, but the avalanche frequency increases with the deforestation, another geologic process).

In many cases we have strong suspicions that man's impact on the landscape has been decisive, but we cannot exclude natural factors. Further research may answer some of the questions, but as I see it our problem is that we do not have a theory for ecological landscape development which is good enough to allow us to formulate our hypotheses in quantitative terms. Testing empirical results against *a-priori* hypotheses is far more satisfactory from the scientific viewpoint than construction of *a-posteriori* hypotheses to explain field observations. It is admitted that a theory for vegetational succession under variable environmental conditions and variable human impact is far beyond the present state of the art. But it would be a great help if forest succession models could be extended to cover a wider range of environmental conditions, and if their accuracy could be improved so that the effects of human impacts could be studied as residuals.

REFERENCES

Carell, B., Forberg, S., Grundelius, E., Henrikson, L., Johnels, A., Lindh, U., Mutvei, H., Olsson, M., Svärdström, K. and Westermark, T. (1987) Can mussel shells reveal environmental history? *AMBIO* **16**, 2–10.

Degens, B.I., Paluska, A. and Eriksson, E. (1976) Rates of soil erosion. In: Svensson, B.H. and Söderlund, R. (Eds), Nitrogen, phosphorus and sulphur—global cycles. SCOPE Report 7. *Ecol. Bull. (Stockholm);* **22**, 185–91.

Douglass, A.E. (1935) Dating Pueblo Bonito and other ruins of the Southwest. National Geographic Society, Contributed Technical Papers. Pueblo Bonito Series, No. 1, Washington, DC.

Drury, W.J. Jr (1956) Bog flats and physiographic processes in the upper Kusokwin River Region, Alaska. Contribution to Gray Herbarium of Harward University, Cambridge, MA.

Fritts, H.C., Smith, D.G. and Stokes, M.A. (1965) The biological model for paleoclimatic interpretation of Mesa Verde tree-ring series. *Am. Antiquity* **31**(2), 101–21.

Jenny, H. (1941) *Factors of Soil Formation*, McGraw-Hill, New York.

Nemec, J. and Rodier, J.A. (1979) Streamflow characteristics in areas of low precipitation (with special reference to low and high flows). In: *The Hydrology of Areas of Low Precipitation*. Proceedings of Canberra Symposium, December 1979, IAHS Publ. No. 128, pp. 125–40.

Polunin, O. and Huxley, A. (1970) *Flowers of the Mediterranean*, Chatto & Windus, London.

Proudfoot, V.B. (1958) Problems of soil history. Podzol development at Goodland and Torr Townlands, Co. Antrim, Northern Ireland. *J. Soil Sci.* **9**, 187–197.

Swanson, P.J., Fredriksen, R.L. and McCorison, P.M. (1982) Material transfer in a western Oregon forested watershed. In: Edmonds, R.L. (Ed.), *Analysis of Coniferous Forest Ecosystems in the Western United States*. U.S./IBP Synth. Ser. No. 14, Stroudsburg, PA, pp. 233–66.

Viereck, L.A. (1970) Forest succession and soil development adjacent to the Chena river in interior Alaska. *Arctic Alpine Res.* **2**, 1–26.

Woodwell, G.M., Hobbie, J.E., Houghton, R.A., Melillo, J.M., Moore, U., Peterson, B.J. and Shaver, G.R. (1983) Global deforestation contribution to atmospheric carbon dioxide. *Science* **222**, 1081–8.

4 Desertification: A Tale of Two Feedbacks

R.D. GRAETZ
CSIRO, Division of Wildlife and Ecology, PO Box 84, Lyneham, ACT 2602, Australia

4.1 INTRODUCTION

4.1.1 DESERTIFICATION: A MAN-MADE PHENOMENON

During the most recent period in the geological history of the earth, changing climate has dramatically transformed the landscapes of its more arid areas. For example in Australia some 20 000 years BP, much of the continent underwent a drying phase with lowered precipitation, prolonged droughts, and higher wind speeds. As a result the extensive dune fields of the core of the continent were rejuvenated (Bowler and Wasson, 1984), the vegetation became more sparse, species composition changed, and some plants and animals became extinct (Kershaw, 1984; Hope, 1984). Similarly, in Africa, there is evidence that, between 9500 years and 4500 years BP, savannahs and grasslands flourished in what is the now the hyper-arid deserts of the eastern Sahara (Ritchie and Haynes, 1987).

Throughout these times, in both locations, modern man inhabited the landscape, experiencing dramatic ecological changes that must have moulded the nature of dependency on the land and the cultural heritage that recorded and perpetuated it.

Desertification is the converse of this process. Man, rather than climate, generates the changes in the landscape. Desertification is the transformation of a landscape from one that did not resemble a desert to one that does. Short-term climate fluctuations, droughts, have usually been associated with the desertification process. However, the evidence supports the argument that the role of climate in desertification is catalytic rather than causal. Desertification is man-made!

4.1.2 DESERTIFICATION AS A DISTURBANCE

Desertification, the degradation or even destruction of arid ecosystems through man's use, is not new. Deleterious changes, measured on the time

Ecosystem Experiments. Edited by H.A. Mooney *et al.*

scale of human archaeological history, have accompanied many civilizations located in the arid and semi-arid areas of the world; for example in China (Hou, 1985) and in the Middle East (Evenari *et al.*, 1985). However, this chapter will concentrate on changes that have occurred over the past few decades.

Desertification is a severe and dramatic disturbance to arid ecosystems. It is named for its transformation of the landscape but its effects can be seen at all trophic levels. For example, during a period of severe landscape degradation in Australia at the end of the last century, the numbers of sheep in the rangelands of New South Wales declined from 13 million in 1890 to 4–5 million in 1900, and since that time has changed very little (Perry, 1968).

Humans are the secondary consumers or top predators in a pastoral system. Sinclair and Fryxell (1985) estimate that earlier this century approximately 50 million nomadic pastoralists used, and were supported by, the Sahelian rangelands. During the epochs of desertification that occurred in the Sahel in the periods 1969–75 and 1984–?, at least one million humans starved, along with 40–50% of the population of domestic stock, and consequently some 7–10 million people became dependent on external food aid (e.g. Wade, 1974; Le Houerou and Gillet, 1986). Similarly the permanent devastation of Australian arid hunter–gatherer systems by the introduction of exotic domestic herbivores and the European rabbit was complete in less than 50 years. Productive systems that had supported man for approximately 2000 generations or more were irreversibly altered by two generations of European pastoralists.

Desertification approaches in extent the global scale of impacts that the 'greenhouse' climatic change may bring. Therefore a disturbance with these attributes should provide some understanding of the structure and functioning of arid ecosystems.

4.1.3 DESERTIFICATION AS AN EXPERIMENT: LIMITATIONS OF THE EVIDENCE

What can be learned from a study of desertification? Can this phenomenon be analysed to increase our understanding of arid ecosystems? Whatever strategy is employed, the task will not be easy because the accounts of desertification suffer from being too closely identified with the human dimensions of the process. The word itself is mostly used in a dramatic, emotional context to maximize attention and focus on the human component to the exclusion of the landscape (e.g. Ahmad and Kassas, 1987).

Desertification is often defined in dynamic terms such as: 'the *expansion* of desert-like conditions and landscapes to areas where they should not occur climatically', e.g. the 'desertization' concept of Le Houerou and Gillet

(1986). While the word desertification implies a process, this term has largely been used to describe the endpoint, i.e. the desertified or degraded landscape.

A simpler and more useful definition is that of Dregne (1983): 'desertification is the impoverishment of terrestrial ecosystems under the impact of man!' Thus desertification is just one form of land degradation that is associated with semi-arid or arid landscapes. Implicit in the above definition is a continuum of ecosystem modification, from slight to severe as the result of the degradation process, and that is the *direct* result of human activity.

Dregne (1983) does not restrict desertification to the semi-arid and arid lands, citing similar outcomes of the degradation process in irrigated and cultivated land. However, in this chapter the concept of desertification is restricted to the semi-arid and arid ecosystems (collectively called 'arid ecosystems') that are grazed by domestic and wild herbivores.

Reports dealing with desertification are numerous, particularly since the UN Conference on Desertification (UNCOD) held in Nairobi in 1977 (United Nations, 1980; Spooner and Mann, 1982; El-Baz and Hassan, 1986). In spite of this abundance, the major difficulty in analysing the phenomenon of desertification is the virtual absence of ecological detail in most reports. In the published work most attention has been paid to describing the higher levels of ecosystem functioning, e.g. the changes in the number of livestock carried, with considerably less emphasis on the reaction of soils, plants, and related processes. Even Sahelian Africa, the site of the most modern epoch of desertification, has not been a source of definitive ecological research on desertification as such. Because international concern has been focused on the human misery, much of the reporting of ecosystem condition and change has been after the event, broad-scale, sweeping, and without the benefit of measurements or assessments of the actual changes that took plaçe.

Substantial ecosystem change has been claimed, but the quantitative evidence for the exact nature and extent of that change is largely lacking. Indeed several authors dispute that the decline in the productivity of some of the Sahel has been permanent or in any way influenced by the state of degradation of the landscape (Hellden, 1986; Olsson, 1985). The application of satellite data to provide objective quantitative estimates of change, e.g. Jacobberger (1987), has little been used.

Given the limitations of the reported evidence, the question of how to use these observations on desertification to better understand ecosystem structure and function becomes difficult. A useful starting point is the word itself. Much of what is known about the modern epoch of desertification in Sahelian Africa comes from the observations of ecologists who have spent their professional lives in the region; see, for example, the comments of Le Houerou and Gillet, 1986. Desertification has been defined in many ways. Verstraete (1986) provides an historical summary of the evolution and use of the word, and indicates how the emphasis on the causal factors (man vs. climate) and the importance of the impacts (soils and vegetation vs economic or social) varies greatly between authors. There is debate whether the process is irreversible or not, progressing or not, and since there is uncertainty about the causes, there is disagreement as to what can be done about it.

The simplest interpretation of the disagreement and disparity between the authors reporting desertification is that the phenomenon is not a simple process. The course and outcome of desertification cannot be unambiguously identified or forecast at any point. Given the very simple nature of the pastoral system, these observations are in keeping with the behaviour of a system wherein the components are very loosely coupled, and that perturbations at one level are amplified by positive feedback at another.

Desertification is derived from 'desert' and the suffix 'fication', which together should be interpreted as the 'making of a desert' or the 'production of desert-like conditions'. This interpretation is of limited ecological utility because degradation of some semi-arid lands, by the encroachment of woody, unpalatable shrubs, results in a wooded landscape very different from that of a typical desert. Also, degradation of arid landscapes may not always result in typical desert landforms, e.g. mobile sand dunes. Lastly, it is difficult to unequivocally substantiate the claim that degradation of arid or semi-arid landscapes always resulted in a 'xerification or desertization' by a permanent reduction in rainfall or rainfall effectiveness. Links between surface conditions (albedo and temperature) and the probability of rainfall have been postulated (Otterman, 1974; Charney *et al.*, 1977) and partly substantiated by modelling (Rasool, 1984; Laval, 1986), but they are disputed by measurement (e.g. Vukovich *et al.*, 1987). Idso (1981) suggests that there is an inadequate understanding of this feedback loop, both of the processes involved and the scaling of their importance from the land surface to the top of the troposphere.

Nonetheless, however varied the definitions of desertification are, the underlying theme is that, depending upon location and initial starting conditions, some of the *ecological*, rather than the *climatological*, characteristics of deserts can be imposed upon the landscape. Ecosystems, through the cumulative impact of man's activities, can approach the structure and functioning of systems that are characteristic of climatically far more arid areas. This is the key issue of this chapter.

4.1.4 STRATEGY OF ANALYSIS

The strategy adopted here is: (1) to review the contemporary understanding of arid ecosystems; (2) to use this theory to explain a case study of desertification.

The theory is tested against the observations of ecosystem structure, processes, and behaviour, particularly at the primary consumer (herbivore) level. The criteria for success in this test will be that observations on the process and outcome of desertification that are available are consistent with contemporary ecological theory.

4.2 NATURE OF ECOLOGICAL THEORY

4.2.1 DESERTIFICATION AS A THEORETICAL PROBLEM?

There is a consensus amongst theoretical and empirical ecologists that ecological systems are intrinsically complex. As an illustration, Holling (1978) has demonstrated the insight gained by locating ecological management problems within the volume defined by the axes of data (empirical knowledge), understanding (theoretical knowledge), and intrinsic complexity. Starfield and Bleloch (1986), in a survey of functional, quantitative approaches to complex ecological problems, reduced the dimensions of this paradigm from three to two, but partitioned this space into four domains to compare and contrast problems and solutions. I have used this paradigm to illustrate the current appreciation of desertification (Fig. 4.1).

Domain 1 of Figure 4.1 is characterized by good data and little understanding. Here statistical/empirical models are most useful in the initial stages of research, e.g. climate change, El Niño. In contrast is domain 4, where there

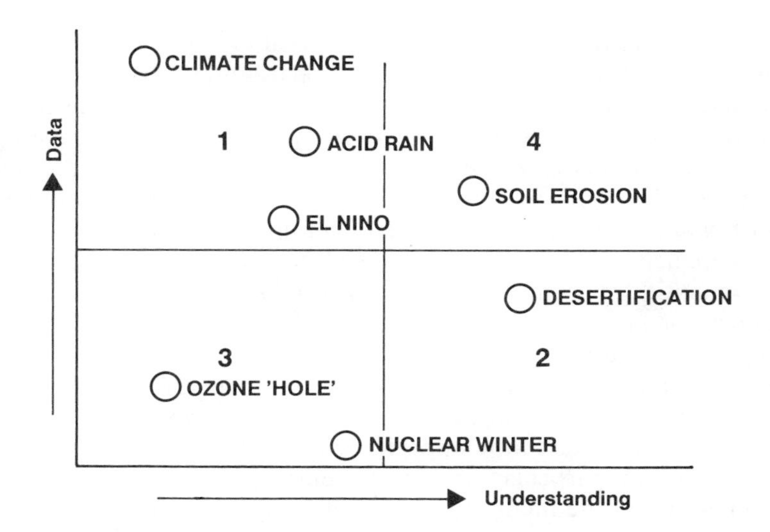

Figure 4.1 A simple illustration of the four domains within the two-dimensional space of understanding (theory) and data. The locations of several global ecological problems are suggested to compare with that of desertification.

are both good data and good understanding. This is the domain of engineering and physics and, consequently, of very few ecological management problems because solutions are readily apparent. I have located desertification in domain 2 because I perceive it to be characterized by good understanding but relatively few data. Desertification is not a problem where the ratelimiting step is ecological understanding. Desertification is not a problem that requires fundamental research. The management of desertification is difficult, not because we do not know what to do, but because we do not know how to do it. In the arid areas, as in the rainforests, the short-term, humanitarian, considerations of people take precedence over scientifically based, long-term solutions for ecosystems.

4.2.2 CONTEMPORARY UNDERSTANDING OF ARID ECOSYSTEMS

In considering the interaction of ecosystems and environment, May (1974) wrote: 'it is to be emphasized that although patterns may underlie the rich and varied tapestry of the natural world, there is no single, simple pattern. Theories must be pluralistic.' The ecological theory available to interpret the pattern and process of arid ecosystems is substantial. The contributions of, particularly, Noy-Meir (1973, 1974a, 1980, 1981, 1985), Shmida *et al.* (1986), Westoby (1980) and Crawford and Gosz (1982) provide a unifying framework that is supported by the voluminous and diverse observations reported in various synthesis volumes (e.g. Goodall and Perry, 1979, 1981; Evenari *et al.*, 1985, 1986).

The most persuasive theory is that of Noy-Meir (1980), who has proposed an 'autecological' hypothesis rather than the 'ecosystem' hypothesis to interpret observations of arid landscapes. The latter implies a system in which there is a high level of connectedness, strong interactions, regulatory or catalytic feedbacks between the components (populations) and the abiotic environment. The organization or structure of such a system, its persistence and functional properties, are distinctive and differ from those of the individual component populations. This model has guided ecologists working with communities that enjoy a high level of constancy or predictability of abiotic environment (e.g. Odum, 1983, 1985; Pimm, 1982; Odum and Biever, 1984; Menge and Sutherland, 1987).

The 'autecological hypothesis' recognizes that, in arid ecosystems, the population dynamics of plants and animals is determined largely by the physical environment and little affected by interactions such as competition and predation. In particular arid ecosystems are water-controlled in an environment where the input of water is highly variable in time, space, and amount. Consequently ecosystem activity is similarly highly variable in time and space. Therefore, in contrast to the 'ecosystem' model, there is a low level of connectedness, weak or intermittent interactions between populations, and few if any regulatory feedbacks.

4.2.3 UTILITY OF ECOLOGICAL THEORY

How do we reconcile or evaluate these two opposing views or theories? The approach taken is pragmatic and utilitarian following Fagerstrom (1987), who reviewed the philosophical and methodological problems facing ecologists required to deal with dynamic systems with a high level of intrinsic complexity.

Ecological theory is a necessary prerequisite for understanding. It is, however, important to differentiate the role or utility of theory from its absolute truth or validity. The ultimate goal of ecological theory is to attain the truth, though it is unlikely this goal will be reached or even recognized. But this should not be the proximate goal of theory. Ecologists do not reject theories because they are wrong. Theories are retained until better ones are developed.

Fagerstrom (1987) proposes four pragmatic criteria by which to judge ecological theory. Theory need not necessarily be true but at least it must be consistent with prevailing ideas. It must be productive of new concepts as well as unify previously disparate ideas. It must be simple in that it is easy to understand the underlying assumptions and to appreciate the implications of those assumptions. Lastly theory should have beauty or appeal. Good theory is attractive when it combines generality with economy and when it is 'universal in content and pregnant in form' (Fagerstrom, 1987).

4.3 NATURE OF ARID ECOSYSTEMS

4.3.1 THE ECOLOGICAL ESSENCE OF ARIDITY

Arid ecosystems are classified in many ways. Most classifications are based on a climatological index of precipitation modified by temperature to produce an index of aridity. This index can be refined further by including measures of the variability and extremes of the key factors, rainfall and temperature (e.g. Shmida *et al.*, 1986). However, it is useful to use the description of Noy-Meir (1973) as a definition. Arid ecosystems are characterized by precipitation so low that it is the dominant controlling factor for most processes.

Precipitation is characteristically highly variable in space and time, occurring as infrequent and discrete events, and this variation has a large stochastic component. There is an emphasis on rainfall sequence in time as well as variability in time and space, because the growing conditions for plants cannot be characterized by averages, even with a measure of variability

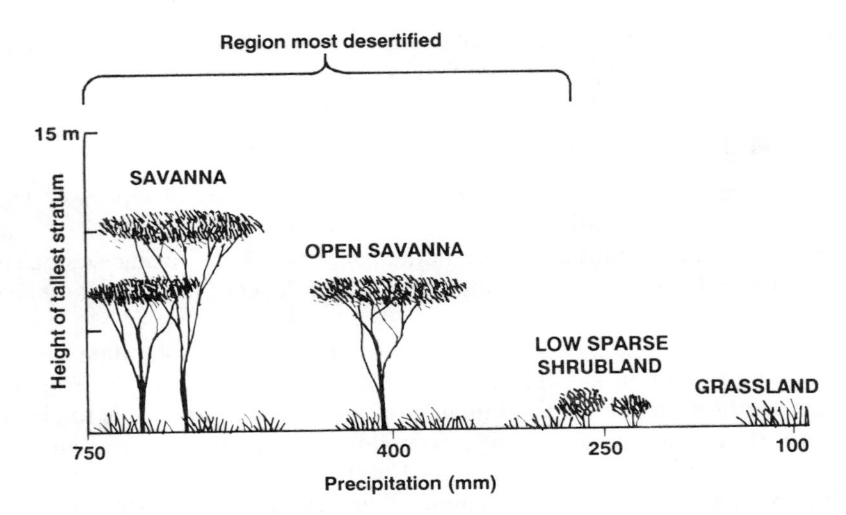

Figure 4.2 A simplified representation of the gradient in structural vegetation types that are here collectively called arid ecosystems. Note that the gradient in rainfall is continuous but non-linear.

attached (Westoby, 1980; Walker, 1987). Different plant growth forms and life strategies have evolved to exploit different rainfall sequences. The viability of a plant functional type or guild in any climate is determined not by an averaged measure of dryness, but by the temporal sequence of favourable growing conditions experienced, and the use of these relative to its competitors (Westoby, 1980).

This definition of arid ecosystems includes not only the systems of the desert margins but also the adjoining semi-arid woodlands or savannahs wherein the availability of soil water also strongly influences the ecological characteristics of these communities (Tinley, 1982; Walker, 1987). In structure the gradient is continuous but non-linear (Fig. 4.2). This wider definition is particularly appropriate because the process of desertification is reported to spread from the high-rainfall woodlands or savannahs to the marginal desert grasslands (Aubreville, 1949; Lamprey, 1983; Sinclair and Fryxell, 1985).

4.3.2 SYSTEM DESCRIPTION

Following Odum (1983, 1985), ecosystems can be described by three elements of (static) structure: biotic and abiotic composition and distribution, and the abiotic environment; and three elements of (dynamic) function: energy flow, nutrient flow, and system characteristics. Not all of these elements are relevant to this analysis of desertification, and a restricted set are discussed below.

Measures of the mean values (high temperature, low precipitation) and their variations are commonly used to illustrate the extreme and variable nature of the physical environment (e.g. Shmida *et al.*, 1986). The variables of temperature and radiation are relatively stable and predictable, in sharp contrast to rainfall which is discontinuous, arriving in discrete pulses. Arid ecosystems also respond in pulses of activity, but the totality of the response is determined not by the total depth of rain but by the time sequence of the independent pulses (Noy-Meir, 1973; Westoby, 1980). Each spike of rain produces a pulse of soil water which, in turn, drives a wave of primary productivity.

The second critical dimension of rainfall variability is spatial. There is a negative correlation between annual precipitation and its variability in time and space. As total rainfall declines its 'patchiness' increases (Noy-Meir, 1973). The spatial heterogeneity in the effectiveness of rainfall in driving a pulse of plant growth is further exaggerated at the landscape level by the factors of edaphic diversity and runoff redistribution. The lower the rainfall, the greater the spatial patterning of vegetation becomes, driven by small differences in topography and soils. The scale of this patterning is of the order of 1000 m or so in the African savannahs (e.g. Tinley, 1982), to 100 m or so in the sparse arid woodlands (e.g. Mabbutt and Fanning, 1987), to 10 m or less in the arid grasslands (e.g. in the African Kalahari—Werger, 1986).

The characteristics of the abiotic environment of arid ecosystems are critical. First because they determine the dimensions of habitat, and habitat is the template of life history strategy acting as a sieve for an organism's dimensions of space (dispersal and foraging range) and (generation) time (Southwood, 1977). Secondly the space/time scaling of habitat, when matched with that of any disturbance, e.g. grazing, fire, erosion, etc., determines the mechanisms of impact of that disturbance.

4.3.3 PRIMARY PRODUCTIVITY

Within arid ecosystems primary productivity and standing biomass are lowest in deserts and highest in the savannah woodlands. Productivity (P) is highly correlated with rainfall (R) and $\mathrm{d}P/\mathrm{d}R$, the marginal productivity, can be used to compare sites (Noy-Meir, 1985; Le Houerou, 1984). A constant $\mathrm{d}P/\mathrm{d}R$ can be interpreted as water limiting plant growth. Where this slope declines indicates that nutrients (particularly N and P) are limiting (Fig. 4.3a). Thereafter, even though biomass and production may be higher, the quality, in terms of N and P content, and therefore the value as a resource for herbivores, is lower. This is of great significance to secondary (herbivore) production, (Fig. 4.3b).

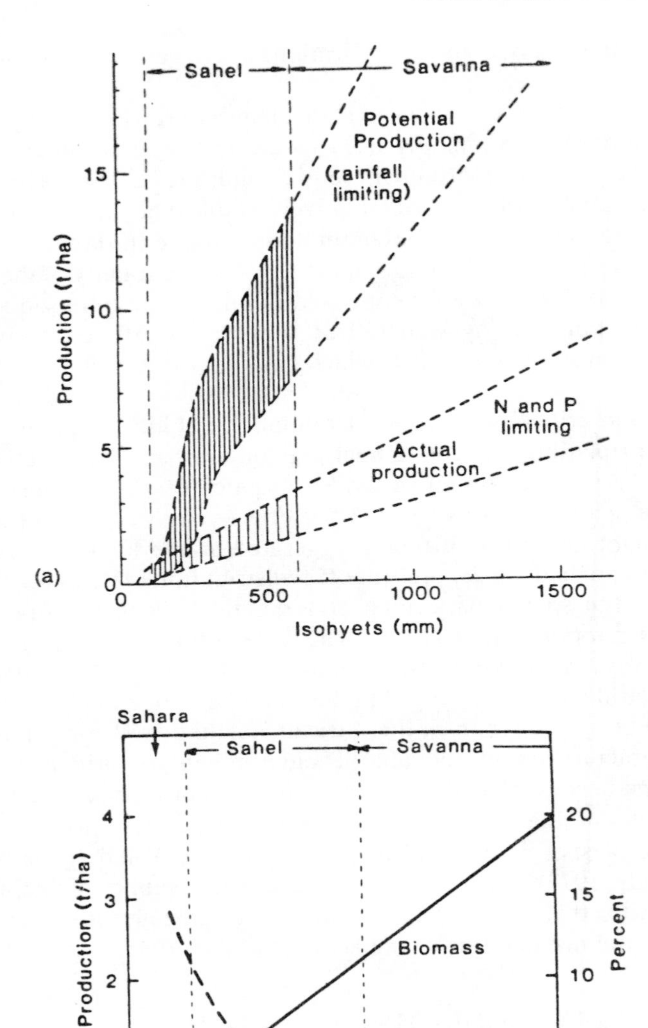

Figure 4.3 (a) The relationship between herbage production and rainfall for African arid ecosystems. The actual values and their range were based on measurements, the potential values were simulated. (b) The relationship between grassland productivity, protein content at September (end of wet season) as a function of mean annual rainfall. Both figures from Breman and de Wit (1983).

The marginal productivity (dP/dR) declines at low values of R and there is usually a threshold of rainfall below which there is no production; i.e. this threshold value defines an effective rainfall. However, small rainfall events often occur as intense showers and are redistributed to run-on areas where production is enhanced well above that had the rainfall not be redistributed. Spatial patterning of vegetation in response to this redistribution of water is common throughout arid ecosystems (Mabbutt and Fanning, 1987; Shmida *et al.*, 1986; Werger, 1986).

The life forms of the vegetation also influence dP/dR. Annual plants have a higher dP/dR than, say, perennial grasses and, particularly, perennial shrubs, which have a large metabolic investment in non-photosynthetic, woody tissues. Conversely perennial plants are able to use the small rainfalls which annuals cannot, and they are able to respond more rapidly to larger, effective rainfalls (Westoby, 1980; Sinclair and Fryxell, 1985; Walker, 1987).

4.3.4 NUTRIENT FLOWS

Although water is the major resource controlling primary productivity, the nutrients N and P particularly exert a strong influence on the functioning of these systems (Breman and de Wit, 1983). There are two critical aspects of nutrients. The first is that nutrients are extracted from large soil volumes by plant roots and then concentrated by litter fall in the vicinity of perennial plants, particularly shrubs. Thus the distribution of general soil fertility (organic matter, nitrogen, phosphorus as well as microbiological activity), which is universally low in the soil at large, becomes spatially and vertically concentrated in the uppermost layers of the soil (e.g. Belsky, 1986; Noy-Meir, 1985; Ruess, 1987). This spatial patterning, 'fertile islands in a sterile desert', may persist long after the plants have been removed (Noble and Tongway, 1986; Walker, 1987).

The second aspect of nutrients is that while turnover is relatively slow it is periodically pulsed by rain. Pool sizes may therefore fluctuate rapidly (Crawford and Gosz, 1982). The combined effect of this spatial and temporal concentration of nutrients is that any disturbance—grazing, fire, or soil erosion—which removes the perennial plants (the nutrient pumps) and redistributes the uppermost (< 5 cm) soil layer can dramatically reduce both the nutrient capital of an ecosystem and, as a consequence, its resilience or capacity to recover from that disturbance (Belsky, 1986; Ruess, 1987).

4.3.5 SECONDARY (HERBIVORE) PRODUCTIVITY

Herbivores (excluding granivores) only utilize a small proportion of available plant production. In more arid environments the pulses of plant growth are unpredictable and transient or, in the wetter savannahs, the quality, rather than the quantity, of the herbage layer is limiting (Breman and de Wit, 1983; Sinclair, 1975). Thus the overall utilization rates of vegetative biomass are low, 5–10% (Noy-Meir, 1985).

The strategies of harvesting this small fraction are critical, however. Considering all trophic levels we can simplify the many dimensions of habitat into just two by the nature of the resources that are offered; 'richness', i.e. quality and quantity, and 'reliability', i.e. frequency and predictability (Shmida *et al.*, 1986; Southwood, 1977). As a consequence of these two habitat dimensions, consumers have evolved two main strategies to use these resources. Populations of short-lived organisms with an intrinsically high rate of increase (insects and rodents) erupt when resources are plentiful and decline when they are not. For long-lived species with a low intrinsic rate of increase (large raptors, mammalian herbivores), the only possible strategy is migration (nomadism) to smooth the variability of resources in space and time.

4.3.6 SYSTEM-LEVEL INTERACTIONS AND FEEDBACKS

Noy-Meir (1973, 1974a, 1980) and Shmida *et al.* (1986) note the lack of evidence for the existence in arid ecosystems of regulatory feedbacks, e.g. competition and predation. However, there is evidence for such feedbacks operating for at least part of the time in the higher rainfall savannahs (Walker, 1987). Nonetheless, the simplest model that underlies all of the characteristics listed above is that arid ecosystems are controlled by water availability and pulses of activity. In very arid environments the pulses may be small and unpredictable, whereas in the higher-rainfall savannahs they are larger and far more predictable (e.g. rainy seasons). These pulses of activity are transmitted through successive trophic levels with little time lag or feedback. The causal network is largely unidirectional. Spatial heterogeneity and variability generated by climate, topography, soils and biota further uncouples potential feedbacks in these systems. Lastly, the abiotic environment is characterized by high variability and extreme values of the significant variables of rainfall and temperature. These extremes can generate changes in species composition that persist long after the event. The perennial components of arid ecosystems act as a memory.

4.4 THE SAHEL OF AFRICA: A CASE STUDY

4.4.1 THE SETTING AND THE SYSTEM

It has been the extensive land degradation, particularly of Sahelian Africa, that has been brought to the attention of the world in the 1960s and again in the early 1980s. Each time the plight of the land and its users was highlighted only because of drought. Indeed, drought is blamed by some not as the catalyst of desertification but as its cause (e.g. Sandford, 1983). The nature of this linkage is convincingly disputed by Sinclair and Fryxell (1985), Courel *et al.* (1984), Rasool (1984) and Gornitz and NASA (1985). The droughts, the prolonged periods of below-average rainfall, served to collapse systems within which there had been slowly building substantial stress as the result of overuse during the between-drought years.

This case study, the description of the processes and effects of desertification on land use in Africa, is primarily based on the writings of Sinclair and Fryxell (1985) and Lamprey (1983). Both authors reject the 'drought hypothesis', i.e. that the two epochs of desertification observed since 1960 are caused by drought; rather they propose an alternative 'overgrazing' hypothesis.

The Sahel of Africa, its name derived from an Arabic word meaning 'the shore', is the southern boundary of the Sahara desert. It is a fringing belt of semi-arid landscape, 5000 km long, stretching from the Atlantic Ocean to the Red Sea. In ecological terms it is a transition zone between the hyper-arid Sahara desert in the north and the humid savannahs in the south. Precipitation varies between 100 and 200 mm in the more arid northern grasslands and 400–600 mm in the southern savannahs. The Sahel is approximately 500 km wide with a rainfall gradient of approximately 1 mm/km (Le Houerou, 1980).

Rainfall in the Sahel results from a continental-scaled weather pattern, the inter-tropical convergence zone (ITCZ), which produces a rainfall belt that moves from south to north. This moving rainfall swathe is followed by migratory insects, birds, and the large mammalian herbivores (Sinclair and Fryxell, 1985). The most spectacular of these migrations are those of the wildebeest (*Connochaetes taurinus*) in the Serengeti of Tanzania and Kenya, and the kob (*Kobus kob*) of Sudan. In both cases more than 1 million animals move from the higher rainfall (600+ mm/year) savannahs, tall perennial grasslands of relatively low quality in the dry season (Jan.–Mar.), to the lower rainfall (< 400 mm/year) but high-quality grasslands in the wet season. These grasslands of the more arid northern edge of the Sahel support a very high animal biomass (density) for a short time while the grass remains green, i.e. resources for the herbivores are rich and plentiful.

Other smaller herbivores that do not migrate en masse have adjusted to regular, seasonally limited supplies of quality food and water by a range of adaptations that include dietary changes (grazing–browsing), or small-scale migration from patch to patch of temporary water and food generated by the spatially patterned rainfalls.

The natural nomadic movement of the wildebeest and kob provides for periods in both wet and dry seasons when there is no, or very little, grazing. During these periods the perennial grasses reproduce and build up reserves to maintain vigour, essential for the persistence of these grasslands under grazing (Belsky, 1986; McNaughton, 1984, 1985; Sinclair and Fryxell, 1985).

A second conclusion is that this nomadic strategy allows a larger population of herbivores to exist than would be possible under a sedentary strategy. Under a sedentary system herbivore populations would be forced to exist on abundant but low-quality grasses for most of the year, and this would reduce growth and reproductive rates. The alternative nomadic strategy provides the opportunity for the harvesting of high-quality forage for a sufficiently long period to improve reproductive success and build reserves to survive on the poor resources of the savannahs during the long dry season (Sinclair *et al.*, 1985).

4.4.2 MAN IN THE SYSTEM

Pastoral societies across the Sahel have until very recently exercised a nomadic strategy very similar to the large native ungulates. Cattle herds moved north into the more arid areas following the rain to graze annual grasses of very high abundance and quality, returning to graze the lower grasslands and agricultural stubble of the sedentary farmers (Breman and de Wit, 1983). Here, as with the large ungulates, the native grasslands and the agricultural lands experienced periods of no grazing.

Sinclair and Fryxell (1985) assemble evidence which suggests that this nomadic system has been in operation for many centuries, possibly 2000–4000 years. The exact behaviour and reaction of this nomadic system under drought is not well known, but it is recognized that at least three major droughts have occurred in the past 100 years as severe as that which, in 1969–75, attracted attention to the Sahelian desertification (Gornitz and NASA, 1985).

Direct and independent support for the importance of nomadism for the persistence of herbivore populations comes from the detailed investigations of energy flow in a Kenyan pastoral ecosystem (Coughenour *et al.*, 1985). They concluded that the key system parameter was the ratio of energy flow to maintenance level requirements and that this ratio must be high to stabilize the system in a variable environment, particularly during the stress of droughts. They concluded that the traditional patterns of operation, which included livestock diversity, nomadism, low energy efficiency and biomass maintenance, promoted the stability and persistence of this pastoral system.

4.4.3 THE CATALYTIC DISTURBANCE

The significance of the 1969–75 drought in the Sahel and the subsequent desertification lies not in its severity but in that it followed upon substantial changes in the nature, size, and behaviour of the nomadic pastoral societies. Sinclair and Fryxell (1985) argue that it was the impact of these social changes that began the desertification process. The drought merely exacerbated and expedited a degradation process that was already in train.

The crucial changes to the pastoral society were precipitated by post-World War II aid which, by its nature and direction, encouraged the expansion of cultivated agriculture in the southern fringes of the Sahel at the expense of the traditional dry season pastures of the pastoralists.

The second aid intervention was the settling of the nomads. Encouraged by the provision of permanent waters, where previously the natural waters had been ephemeral, and/or enforced by new nationalist governments, nomadism declined dramatically in the early 1960s.

Cultivation decreased the resources available to the pastoralist and settlement enforced sedentary, year-long grazing of grasslands by a population of humans and cattle that grew rapidly (Brown and Wolf, 1984). This rapid growth in the populations of both pastoralists and cattle, $\approx$ 3% per year, resulted from the removal of a previously limiting population control, the high mortality rates of the newborn. This control on both human and livestock populations was suppressed by access to medical and veterinary services following settlement. Lamprey (1983) has summarized the intricate social and ecological interaction that develops, and how all increased human activity is transmitted to, and focused on, the land.

The net result of these initial, catalytic social changes has been the overgrazing of the landscape. The cumulative impact of the unrelenting overgrazing, in concert with the additional stress of drought, was desertification, the degradation of landscapes over extensive areas. Ironically, where bores had been sunk by aid projects to supply permanent water and remove a perceived limitation of ephemeral supplies, large numbers of cattle died of hunger, whereas before they had only rarely died of thirst (Wolf, 1986).

It is this changing of the nature of the landscape, the removal of more vegetation by grazing animals and humans than the system requires for it to function without change of state, and the initiation and enhancement of soil erosion, that encapsulates desertification. That is, the landscape is transformed to resemble one far less productive, and therefore by association, far more arid, than the climate would otherwise determine (Verstraete, 1986).

4.4.4 THE PHENOMENON OF DESERTIFICATION

To explore this transformation requires a shift in focus from the primary consumers to the primary producers and associated landscape processes. Few detailed studies at this level have been reported specifically for desertification in the Sahel. Under the assumption that these processes are not geographically unique, a generalized account has been produced based on Warren and Maizels (1977) and Reining (1978), and supplemented by research from elsewhere.

The direct effects of grazing are on the yield (standing crop) of the herbage (grass and forb) and browse (woody perennial) layer. Depending upon life form, grazing affects population dynamics of the species. Overgrazing can completely eliminate perennial—but may have little effect on annual—plants (see Noy-Meir, 1975). The spread of the grazing impact is determined by the landscape heterogeneity, the herbivore type, and the time of year. Separation of, or competition among, herbivore species is determined by these three dimensions—location, forage type, and time. Diet preferences are characteristically different among herbivore species. Diets change with time as either the relative abundance or quality, or both, of the grazed plant species changes with time; see for example the diet studies published by Coppock *et al.* (1986a) and Moore (1987).

A reciprocal interaction exists between the vegetation and the herbivore. The quality of the diet in terms of energy and nutrient content is determined by the selective capability of the herbivore, but is set by the base level offered by the forage (Coppock *et al.*, 1986b; Sinclair, 1975; Skarpe and Bergstrom, 1986). In these environments highly opportunistic consumers appear to enjoy an advantage (Coppock *et al.*, 1986a; Moore, 1987).

How much of the standing crop of vegetation can be harvested without seriously affecting the future productive capacity of the grass sward is the central question of range management. In a seasonal or stochastic rainfall environment the productivity can be expected in pulses. A nomadic system as described above would harvest intensively for a short time, usually close to the peak of productivity, and depart before the pulse of soil water was depleted. Thus grass growth would continue, root reserves be replenished and reproduction take place without the stress of grazing.

In variable and unpredictable ecosystems the level of harvesting by long-lived herbivores must be low for populations to be sustained. The more variable the rainfall–pasture growth cycle, the lower the overall level of offtake (Caughley *et al.*, 1987). A harvesting pattern under nomadic grazing is intermittent and varies according to seasonal conditions. The feedback controlling the herbivore usage is primarily the availability of drinking water and secondarily the biomass levels. The two will be positively correlated and act in concert; high-rainfall years have abundant casual drinking water, whereas in dry years animals are forced to migrate elsewhere, presumably before the level of offtake becomes deleterious. This model is simplistic and ignores the interactions that occur at plant level between defoliation and subsequent growth, which will be discussed later.

The most influential long-term consequence of sustained and intense grazing on the productivity of the herbage layer is indirect in action. It is affected through altering the patterns of soil water and nutrient flow in the vicinity of plants and results from the cumulative physical impact of the herbivores on the partitioning and redistribution of rainfall. Both nutrient cycling and water flow are patterned in a scale of size of individual plants (Noy-Meir, 1975; Noble and Tongway, 1986). The removal of fallen dead plant material (litter) by grazing and (mostly) trampling changes the partitioning of rainfall at the soil surface. As the surface is bared, infiltration is decreased and runoff and redistribution increased. This change is first catalysed by the removal of litter which has served to reduce the kinetic energy of raindrop impact and facilitate infiltration close to the perennial plants (Braunack and Walker, 1985; Noble and Tongway, 1986).

The removal of litter and exposure of the soil surface initiates two positive feedback loops. The first is that baring the soil enhances the formation of soil crusts by direct impact of raindrops. These crusted soils immediately begin to reduce infiltration and encourage redistribution of rainfall. In the longer term the reduced soil water regimes will reduce plant growth, and therefore cover, over the soil and the positive feedback loop is established.

The loss of litter also facilitates the erosion and redistribution of soil by wind (Warren and Maizels, 1977; Noble and Tongway, 1986). This is just as effective as water erosion in dramatically altering the spatial patterns of soil water and nutrient flow. Like water erosion it can also be a positive feedback process because of the non-linear relationships, the erosive power of either agent and the relative proportion of protective plant cover. Very small changes in plant cover at or about the threshold, catalyse disproportionately larger changes in erosion (Fig. 4.4).

The second exacerbating loop is the change in microclimate that comes with reducing plant cover. The greater the reduction in plant cover, the more hostile to plant growth and establishment the physical environment becomes. In particular the temperature and evaporative regimes experienced by plants, especially seedlings, can be dramatically amplified by increasing the area of bare soil in the sward. The outcome of this feedback has often been observed in fenceline contrasts where ungrazed grasslands remain greener longer than do grazed ones. It has been observed at many scales, from the small exclosures of plant research (e.g. Belsky, 1987) to the large areas that are directly observable by satellite (e.g. Wade, 1974; Otterman, 1981; Otterman and Tucker, 1985).

The end-result of these processes is desertification. The positive feedback loops continue to enhance runoff and redistribution of water at the expense of infiltration. This runoff carries with it both soil and nutrients, and seals the surface over which it runs, further adding to the positive feedback loop. Where the nutrients and water are redistributed depends upon the characteristics of both landscape and rainfall. Redistribution may exaggerate or rescale existing spatial pattern. For example, where water was mostly redistributed over the scale of inter-plant distance (0.1 m) in stable communities, it may now be redistributed over the larger scales of erosion cells (Pickup, 1985) or the clumping or banding of vegetation (Foran, 1987; Mabbutt and Fan-

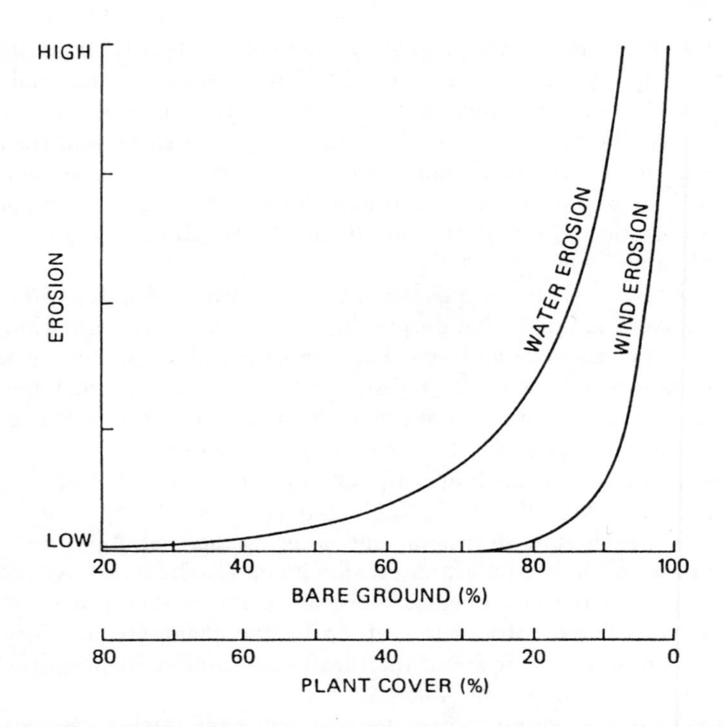

Figure 4.4 The relationship between the erosion potential of water and wind as a function of the relative proportion of plant cover. From Marshall (1973).

ning, 1987). Ultimately the redistribution of water and nutrients from the most frequently experienced rainfalls may routinely reach the scales of landscape drainage representing an export of water and nutrients, both of which critically control the productivity of arid ecosystems.

These positive feedback processes result in a transformed landscape. An appealing metaphor is an unravelling of a fabric, the fabric of ecological relationships, and the pattern of the landscape frays (Wolf, 1986). As the scale of patterning of the system increases, the remnant pockets of vegetation begin to function independently of one another and the level of connectedness and interaction within the system declines.

4.5 A SYSTEMS ANALYSIS OF DESERTIFICATION

4.5.1 COMPONENTS AND INTERACTIONS

Man and arid ecosystems can be combined as a simple four-component system; man, and society, herbivores, herbage and soils, and climate (Fig.

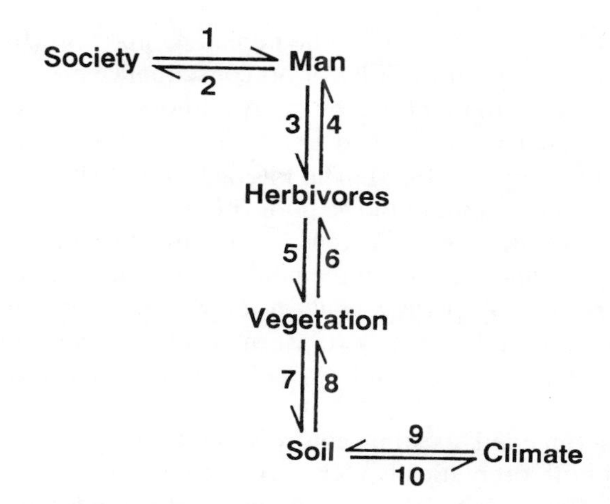

Figure 4.5 A generalized model of the structure of a Sahelian nomadic pastoral system. The ecosystem is represented by just four components: man, herbivores, vegetation, and soils, and two influences: climate and society. The ten reciprocal interactions individually comprise several ecosystem processes, e.g. herbivory, infiltration, predation. Note that the system is unidirectional (unbranched).

4.5). In this system there are 10 interactive links between components that vary in strength. Each of these links may involve one or more ecological processes, e.g. herbivory or predation, and depending upon context the paired links operate as either positive or negative feedback loops.

The proposition is that, within this simple system model, desertification results from changes in the strength and sign of the interactive links. No new patterns of linkage are established. The transformation is effected by a change in the nature of two feedbacks, from weak to strong and from negative to positive. That is, the beginning process of desertification will, by positive feedback, increase the probability that further desertification will occur.

This is not a new proposition. It is a restating of commonly expressed conclusions that in desertification cause and effect are difficult to separate (Reining, 1978), or that 'desertification feeds on itself' (UNCOD, 1977; Warren and Maizels, 1977). This proposition will be supported by examining each of the interactions involved.

4.5.2 SOCIETY AND PASTORAL MAN; INTERACTIONS (1) AND (2)

These two interactions are critical to the process of desertification. It is proposed that (1) was the trigger for the contemporary epoch of desertification in Sahelian Africa and that (2) is the only way this problem can be managed in the future. Desertification is a man-made process. Therefore the solution can only be effected through the human component. Intervention at any lower level in this system will be of no consequence.

Interaction (1) has been identified by many authors as the primary cause of desertification in Sahelian Africa and elsewhere, (Noy-Meir, 1974b; Sinclair and Fryxell, 1985; Lamprey, 1983). The interaction in Sahelian Africa included the reduction in the amount of the poor (but reliable) resources available to the nomadic pastoralists by the extension of cultivation into the dry-season savannah grasslands, and the encouragement of the transition from a nomadic to a sedentary system of grazing. The latter was most important because contact with non-arid and urban systems, medical care, veterinary services, and even supplementary feeding for livestock, became available for the first time.

These factors removed existing controls on both human and livestock populations and both dramatically increased, exceeding the carrying capacity of the system. The 'crash', the beginning of the landscape degradation process, was exacerbated by the additional stress of drought. But drought was not the primary causal agent. The primary causal agent was the greatly increased impact of herbivores, interaction (5), on the vegetation and soils, i.e. 'overgrazing'. This overgrazing was precipitated by (i) an absolute increase in the population of pastoralists ($\approx 3\%$ per year) and domestic stock, too many in time, and (ii) by the transition to sedentary grazing, too many in space.

In the history of mankind, nomadic pastoralism was once the pursuit of an appreciable proportion of the world's population, particularly within the savannah biome (Hadley, 1985). Today they are a minority, enjoying little social prestige or political power (Spooner and Mann, 1982; Sandford, 1983). In Fig. 4.5 the interaction (2) is also strong. The future of pastoralists, pastoralism (and arid ecosystems?) is being determined by socio-political forces from outside the arid lands. Interaction (2) is currently a positive feedback. The plight of the pastoralists in the 1969–75 drought has increased the flow of aid and other support to maintain the existing levels of human population which nullified the negative feedback of diminished food supply. The result of this has been that after the 1969–75 epoch was an even more severe one in 1980–84 (Breman and de Wit, 1983; Sinclair and Fryxell, 1985). The long-term destructive effect of this positive feedback has been recognized, and has led to recommendations that aid be tied to land-management change to reinstate some system control (e.g. Breman and de Wit, 1984; Sinclair and Fryxell, 1985).

4.5.3 MAN AND HERBIVORES; INTERACTIONS (3) AND (4)

Pastoral man interacts with domestic herbivores (3) by control of populations (species and numbers) in time and their disposition in space. Various management strategies have evolved over the millennia shaped by the characteristic of the physical environment. Nomadism, in the example of the Sahel, was cyclic–seasonal following the rains. In more arid, less predictable environments, nomadism is often completely opportunistic. The objective of nomadism is to harvest the 'rich' (high quantity and quality) resources which, in the Sahel, were also reasonably reliable in time (rainy seasons) and space (the movement of the ITCZ). The selection criteria for interaction (3) is survival and the historical evidence is that these pastoral societies and systems have been remarkably resilient (Coughenour *et al.*, 1985; Hadley, 1985; Lamprey, 1983; Sinclair and Fryxell, 1985).

No doubt there was variation in numbers with climate, but the system embodied sufficient resilience to persist, i.e. the system was stable (Holling, 1973; Noy-Meir, 1974b; Walker and Noy-Meir, 1982). Such stability can occur only if there are regulating interactions or negative feedbacks that tend to reduce the forward reaction. Interaction (4) is one of the negative feedbacks operating within grazed systems and ecosystems. Herbivores are harvested to provide meat, milk, and blood. These three items can comprise > 70% of the energy available to pastoral man, with milk usually being the largest component (Coughenour *et al.*, 1985). Because herbivore products provide almost the total resource harvested, and these products are energetically rich, this provision represents a considerable demand on the herbivore population. Even in a system which is maintenance- rather than production-oriented, this is a limiting interaction. There is a limited number of livestock that can be safely harvested given low reproductive rates and the variability of climate. The rates of increase of domestic stock are considerably less than that of the vegetation, and herbivore density cannot closely track the biomass of pasture. Because the grazer and the grazed operate on different scales of time they are not closely coupled. This 'slack' in the system has been reported for other grazing systems (e.g. Caughley *et al.*, 1987) and is associated with stability or persistence.

4.5.4 HERBIVORES AND VEGETATION; INTERACTIONS (5) AND (6)

Herbivores harvest vegetation, interaction (5). The distribution of this harvest across vegetation types differs with herbivore species and for any one species with time (e.g. Noy-Meir, 1974a; Coppock *et al.*, 1986a,b). It can be generalized that herbivores, ruminants or not, seek the 'rich' resources, herbage that is highest in quality (digestibility, N, etc.). When faced with 'poor' resources, i.e. herbage of low quality, herbivores, particularly domestic stock, show considerable selectivity (finding green plant parts) or flexibility (switching plant types) in their diets (Breman and de Wit, 1984; Sinclair and Fryxell, 1985; Coppock *et al.*, 1986a; Moore, 1987) with the objective of maximizing the nutritional value of the intake (Coppock *et al.*, 1986b).

In persistent, arid pastoral systems the efficiency of harvest is low (Noy-Meir, 1974a, 1985), as is the efficiency of conversion (Coughenour *et al.*, 1985). In destabilized or desertifying systems this harvest rate can obviously reach 100% where the landscape is completely denuded of vegetation. However, the utilization rate does not have to reach that level to have a significant effect on the herbage layer. There is a vast literature describing the direct effects of repeated defoliation on the productivity and, to a much lesser extent, the longevity of perennial grasses and shrubs. The timing and cycle of defoliation in relation to the phenological stage of the plant is the critical factor (Hodgkinson and Mott, 1987). High levels of offtake involve risk. The greater the offtake, the longer is the recovery time because recovery time is a function of existing productive capacity. A long recovery time makes the perennial grassland system more vulnerable and less resilient, because of the increased probability of a drought intervening and generating additional stress.

At lower levels of grazing there is a synergistic reaction of plant growth to grazing. The higher the plant biomass, the smaller the growth response for a given increment of rainfall. Thus a reduction of biomass by grazing can enhance the productivity of the vegetation as a whole (Noy-Meir, 1975). The existence of this synergistic process has been proposed by McNaughton (1984), and the importance of it explored by Caughley *et al.* (1987).

The interaction of vegetation on herbivores (6), serves as a negative feedback to grazing (5). Setting aside adaptations such as spinescence, toxins, etc., the predominant factors in this negative feedback to the herbivore are limitations to the supply of energy and nutrients (Ruess, 1987 and references therein). In the African savannahs and grasslands the considerable range in species composition, each with its distinct phenophase, has resulted in a remarkable species packing of herbivores. Each herbivore occupies a tightly defined resource niche, each having different nutritional requirements.

As the vegetation complex declines in 'richness' or quality, herbivores can respond by migration on some spatial scale. Where this is not possible, the response must be in diet selectivity and flexibility. Diets low in quality (energy and nutrients) reduce individual vigour and reproductive success, thereby providing a negative feedback on herbivore numbers (Sinclair *et al.*, 1985).

The strength of the reciprocal coupling of mammalian herbivores to vegetation varies within arid ecosystems, being stronger in the higher-rainfall savannahs than in the lower-rainfall grasslands. At the arid extremes there is almost no coupling (Shmida *et al.*, 1986). The strength of this coupling is determined not by the variability in the quality and quantity of the vegetation but by its predictability in time and space. Stochastic systems can only be utilized by sedentary herbivores with either short reproductive times (e.g. insects) or the capacity to store forage (e.g. termites—Braithwaite *et al.*, 1988) or by opportunistic nomadism. Where predictability is high, i.e. pronounced and reliable rainy seasons, then this coupling of herbivores to vegetation by energy and nutrient flow can be tight and complex, and has been interpreted as stabilizing the population of both (McNaughton, 1985; Belsky, 1986).

4.5.5 VEGETATION AND SOILS; INTERACTIONS (7) AND (8)

The physical and chemical characteristics of soils largely determine the control of vegetation composition and abundance by the climatic factor of rainfall, interaction (8). Vegetation, however, is not completely passive. Plants create a physical and chemical microenvironment. Interaction (7) includes the processes by which perennial vegetation determines the extent and spatial pattern of abundance and flow of water and nutrients in the landscape (Noble and Tongway, 1986). This concentration and patterning facilitates the perpetuation of the species in space and therefore leads to a persistence of the community in space and time.

Interaction (8) serves as a positive feeback to (7). An enhanced concentration of water and nutrients in turn enhances the establishment, growth and persistence of vegetation. The control of this positive feedback loop is the absolute levels of, and variability in, the resources of water and nutrients. It is one of the most influential and stabilizing of feedback loops in arid ecosystems.

4.5.6 CLIMATE AND SOILS; INTERACTIONS (9) AND (10)

The climate factor of rainfall drives arid ecosystems, interaction (9). This is effected through a hierarchy of influences. At one extreme is the macroclimate which, by virtue of scale, is larger than arid ecosystem and thus outside its influence. At the other extreme is the microclimate or microenvironment, which is determined by the vegetation (Noy-Meir, 1980).

The strength and sign of the feedback (10) remains uncertain. The nature of the landscape surface undoubtedly influences the characteristics of the atmosphere passing over it. It has been proposed that this link can, under conditions of desertification, operate as a positive feedback (see section 4.1.3). The relative strength and scale of this influence compared with those of the mesoscale or global circulation mechanisms is not clear. For example the two periods of continent-wide drought associated with the modern epochs of desertification in the Sahel of Africa are now known to have been associated with the El Niño oceanic circulation anomaly of the Pacific Ocean.

4.6 SUMMARY AND CONCLUSIONS

Desertification was first named forty years ago to describe the destructive transformation of savannah ecosystems in Africa. 'These are real deserts that are being born today, under our eyes, in the regions where the annual rainfall is from 700–1500 mm' (Aubreville, 1949, translated in Glantz and Orlovsky, 1983). Twenty years later the process became a large-scale reality; it was swift, extensive, and severe.

Within the catalogue of ecosystem perturbations desertification could be classified as a 'pulse' experiment; a rapid and severe alteration to the system (Bender *et al.*, 1984). Such experiments theoretically can yield information about the direct interactions between ecosystem components as the system 'relaxes' back to equilibrium. However, the actual difficulties in achieving this goal are formidable (Bender *et al.*, 1984; Yodzis, 1988).

An equilibrium-centred view is not appropriate for arid ecosystems. Desertification appears to have been an irreversible perturbation. The landscapes of the Sahel now operate within a new domain of stability with attributes and behaviour more closely akin to ecosystems typical of more arid climates.

Within the limitations of the evidence available, and admittedly with much inference and generalization, it has been possible to construct a feasible explanation of the ecological processes involved in desertification using a very simple model of grazed arid ecosystems (Fig. 4.5). The model has just four components (soils, herbaceous vegetation, herbivores, and man), and two external influences (external society and climate). They are connected by ten interactions, each of which may comprise several processes, e.g. herbivory, infiltration. The pairs of interactions may function as loops and be either strong or weak, positive or negative. In terms of dynamic food chain links this system would be ranked as 2 (Fretwell, 1987).

Given the recorded sequences of events in the Sahel, this model was adequate to explain the observed consequences of those events.

The nomadic pastoral system has persisted for a considerable time experiencing droughts known to have been as severe as those of 1969–75. Therefore we may conclude that this system had evolved as a stable, resilient strategy. This conclusion is supported by the similarity in grazing strategy between the nomadic pastoralism and the large migratory herbivores, the wildebeest and the kob.

In its undisturbed state the system was driven by climate (rainfall), interaction (9), and controlled by negative feedbacks between vegetation and herbivores, interaction (6), and to a lesser extent between herbivores and man, interaction (4). In the former, the negative nature of the feedback was determined by the nutritional limits to herbivore health, growth, and reproduction offered by the low-quality but reliable dry-season grasslands. The grasslands acted as a 'passively harsh' environment for herbivores (Oksanen, 1988). In the latter case the obligate but hazardous dependence of the pastoralist on the herbivores must also be seen as a limit to human populations through similar controls of disease, growth, and reproduction.

The patterns of nomadic movements of the livestock can be interpreted as a strategy to harvest the higher-quality, but less predictable, arid grasslands. This nomadic strategy incorporated an additional control in the form of availability of water for livestock. Its availability determined the distribution and density of herbivores sympathetically with the productivity of the grasslands.

Consistent with this model the modern epoch of desertification can be explained by one small but catalytic change. It was that the isolation, and therefore the integrity, of the nomadic pastoral society was breached by the interaction with alien economic systems starting about 1950. This produced two responses. First a pulse of population growth of pastoralists and (then) livestock passed from the higher levels to the lower. Secondly a controlling negative feedback between vegetation and herbivores, lack of drinking water, was overridden by the supply of permanent water. This allowed the pulse of disturbance passing down the system to be transmitted to the vegetation and soils where before it would have been suppresssed.

Under additional stress of the first drought (1969–75), the interaction exceeded some threshold whereby the interaction (8), previously weak and positive, was transformed into strong negative feedback on plant growth and establishment by soil erosion by wind and water. This continued in a self-reinforcing way until a new domain of resilience had been reached. Most interactions had been spatially uncoupled, productivity at all levels had declined, become episodic, and the landscape looked like a desert.

The collapse of the primary productive system, i.e. diminished production, was transmitted back up the system (1975–?) and interaction (2) became a strong positive feedback on the further aid intervention (1). The cycle repeated itself by 1980–? Given the continual input of ecologically inappropriate intervention, the cycle will continue at an every-increasing frequency.

In summary, desertification is a tale of two feedbacks. A nomadic pastoral system was stabilized by the negative feedbacks between the grazed, the grazer, and the grazier. The ecological processes that comprised these feedbacks, e.g. energy and nutrient flow, were determined by the characteristics of arid ecosystems wherein the influence of space and time have exaggerated importance.

The nomadic system of the Sahel was destabilized by the effective removal of two negative controlling feedbacks. The resultant pulse of disturbance and the drought precipitated a negative feedback of soil erosion which desertified the productive base of the system. This pulse of response, a new and severe negative feedback, passed back up the system only to solicit a further positive feedback of destabilizing aid to renew the cycle again.

Desertification as an ecological phenomenon reinforces the view that, for arid ecosystems, abiotic factors are dominant. Simple unidirectional representations of ecosystems are consistent, adequate, and intuitively appealing.

REFERENCES

Ahmad, Y.J. and Kassas, M. (Eds), (1987) *Desertification: Financial Support for the Biosphere*, Hodder and Stoughton, London.

Aubreville, A. (1949) *Climats, Forets et Desertification de l'Afrique tropicale*, Societé d'éditions Geographiques, Maritimes et Coloniales, Paris.

Belsky, A.J. (1986) Population and community processes in a mosaic grassland in the Serengeti, Tanzania. *J. Ecol.* **74**, 841–56.

Belsky, A.J. (1987) The effects of grazing: confounding of ecosystem, community, and organism scales. *Am. Nat.* **129**, 777–83.

Bender, E.A., Care, T.J. and Gilpin, M.E. (1984) Perturbation experiments in community ecology: theory and practice. *Ecology* **65**, 1–13.

Bowler, J.M. and Wasson, R.J. (1984) Glacial age environments of inland Australia. In: Vogel, J.C. (Ed.), *Late Cenozoic Paleo Climates of the Southern Hemisphere,* Balkema, pp. 183–208.

Braithwaite, R.W., Muller, L. and Wood, J.T. (1988) The ecological structure of termites in the Australian tropics. *Aust. J. Ecol.* **13** (in press).

Braunack, M.V. and Walker, J. (1985) Recovery of some surface soil properties of ecological interest after sheep grazing in a semi-arid woodland. *Aust. J. Ecol.* **10**, 451–60.

Breman, H. and de Wit, C.T. (1983) Rangeland productivity and exploitation in the Sahel. *Science* **221**, 1341–7.

Brown, L.R. and Wolf, E. (1984) Food crisis in Africa. *Nat. Hist.* **93**, 16–20.

Caughley, G., Shepard, N. and Short, J. (1987) *Kangaroos: Their Ecology and Management in the Sheep Rangelands of Australia*, Cambridge University Press, Cambridge.

Charney, J.G., Quirk, W.J., Chow, S. and Kornfield, J. (1977) A comparative study of the effects of albedo change on drought in the semi-arid regions. *J. Atmos. Sci.* **34**, 1366–85.

Coppock, D.L., Ellis, J.E. and Swift, D.M. (1986a) Livestock feeding ecology and resource utilization in a nomadic pastoral ecosystem. *J. Appl. Ecol.* **23**, 573–83.

Coppock, D.L., Swift, D.M. and Ellis, J.E. (1986b) Seasonal nutritional characteristics of livestock diets in a nomadic pastoral ecosystem. *J. Appl. Ecol.* **23**, 585–95.

Coughenour, M.B., Ellis, J.E., Swift, M.D., Coppock, D.L., Galvin, K., McCabe, J.T. and Hart, T.C. (1985) Energy use and extraction in a nomadic pastoral ecosystem. *Science* **230**, 619–25.

Courel, M.F., Kandel, R.S. and Rasool, S.K. (1984) Surface albedo and the Sahel drought. *Nature* **307**, 528–31.

Crawford, C.S. and Gosz, J.R. (1982) Desert ecosystems: their resources in space and time. *Environ. Conserv.* **9**, 181–95.

Dregne, H.E. (1983) *Desertification of Arid Lands*, Harwood, New York.

El-Baz, F. and Hassan, M.H.A. (Eds) (1986) *Physics of Desertification*, Martinus Nijhoff, Dordrecht.

Evenari, M., Noy-Meir, I. and Goodall, D.W. (Eds) (1985) *Ecosystems of the World*, vol. 12A: *Hot Deserts and Arid Shrublands*, Elsevier, Amsterdam.

Evanari, M., Noy-Meir, I. and Goodall, D.W. (Eds) (1986) *Ecosystems of the World,* vol. 12B: *Hot Desert and Arid Shrublands*, Elsevier, Amsterdam.

Fagerstrom, T. (1987) On theory, data and mathematics in ecology. *Oikos* **50**, 258–61.

Foran, B.D. (1987) Detection of yearly cover change with Landsat MSS on pastoral landscapes in Central Australia. *Remote Sens. Environ.* **23**, 333–50.

Fretwell, S.D. (1987) Food chain dynamics: the central theory of ecology. *Oikos* **50**, 291–301.

Glantz, M.H. and Orlovsky, N. (1983) Desertification: a review of the concept. *Desert. Control Bull.* **9**, 15–22.
Goodall, D.W. and Perry, R.A. (Eds) (1979) *Arid Land Ecosystems*, vol. 1, Cambridge University Press, Cambridge.
Goodall, D.W. and Perry, R.A. (Eds) (1981) *Arid Land Ecosystems*, vol. 2, Cambridge University Press, Cambridge.
Gornitz, V. and NASA (1985) A survey of anthropogenic vegetation changes in West Africa during the last century—climatic implications. *Climatic Change* **7**, 285–325.
Hadley, M. (1985) Comparative aspects of, and use and resource management in, savannah environments. In: Tothill, J. and Mott, J.J. (Eds), *Ecology and Management of the World's Savannahs*, Australian Academy of Science, Canberra, pp. 142–58.
Hellden, U. (1986) Desertification monitoring: remotely sensed data for drought impact studies in the Sudan. *Proceedings ISLSCP Conference, Rome, Italy*, European Space Agency Publication SP-248, Paris, France, pp. 417–28.
Hodgkinson, K.C. and Mott, J.J. (1987) On coping with grazing. In: Horn, F.P., Hodgson, J., Mott, J.J. and Brougham, R.W. (Eds), *Grazing-Lands Research at the Plant–Animal Interface*, Winrock International, Morrilton, AK, pp. 171–92.
Holling, C.H.S. (1973) Resilience and stability of ecological systems. *Ann. Rev. Ecol. Syst.* **4**, 1–23.
Holling, C.H.S. (1978) *Adaptive Environmental Assessment and Management*, John Wiley & Sons, Chichester.
Hope, G. (1984) Australian environmental change: timing, directions, magnitudes and rates. In: Martin, P.S. and Klein, R.G. (Eds), *Pleistocene Extinctions: Prehistoric Revolution*, University of Arizona Press, Tucson, AZ, pp. 681–9.
Hou, Ren-zhi (1985) Ancient city ruins in the deserts of the Inner Mongolian Autonomous Region of China. *J. Hist. Geog.* **11**, 241–52.
Idso, S.B. (1981) Surface energy balance of the deserts and the genesis of deserts. *Arch. Met. Geoph. Biokl., Ser. A*, **30**, 253–60.
Jacobberger, P.A. (1987) Geomorphology of the upper Inland Niger Delta. *J. Arid Environ.* **13**, 95–112.
Kershaw, A.P. (1984) Late Cenozoic plant extinctions in Australia. In: Martin, P.S. and Klein, R.G. (Eds), *Quaternary Extinctions: a Prehistoric Revolution*, University of Arizona Press, Tucson, AZ, pp. 691–707.
Lamprey, H.F. (1983) Pastoralism yesterday and today: the overgrazing problem. In: Bourliere, F. (Ed.), *Tropical Savannahs*, Elsevier, Amsterdam, pp. 643–66.
Laval, K. (1986) General circulation model experiments with surface albedo changes. *Climatic Change* **9**, 91–102.
Le Houerou, H.N. (1980) The rangelands of the Sahel. *J. Range. Manage.* **33**, 41–6.
Le Houerou, H.N. (1984) Rain use efficiency: a unifying concept in arid-land ecology. *J. Arid Environ.* **7**, 213–47.
Le Houerou, H.N. and Gillet, H. (1986) Desertization in African arid lands. In: Soulé, M.E. (Ed.), *Conservation Biology: The Science of Scarcity and Diversity*, Sinauer, Sunderland, MA, pp. 444–61.
Mabbutt, J.A. and Fanning, P.C. (1987) Vegetation banding in arid Western Australia. *J. Arid Environ.* **12**, 41–59.
McNaughton, S.J. (1984) Grazing lawns: animals in herds, plant form, and coevolution. *Am. Nat.* **124**, 863–886.
McNaughton, S.J. (1985) Ecology of a grazing ecosystem: the Serengeti. *Ecol. Monographs* **55**, 259–94.
Marshall, J.K. (1973) Drought, land use and soil erosion. In: Lovett, J. (Ed.), *Drought,* Angus & Robertson, Sydney, pp. 55–80.

May, R.M. (1974) Ecosystem patterns in randomly fluctuating environments. In: Rosen, R. and Snell, F. (Eds), *Progress in Theoretical Biology*, Academic Press, New York, pp. 1–50.

Menge, B.A. and Sutherland, J.P. (1987) Community regulation: variation in disturbance, competition, and predation in relation to environmental stress and recruitment. *Am. Nat.* **130**, 730–57.

Moore, P.D. (1987) Mobile resources for survival. *Nature* **324**, 198.

Noble, J.C. and Tongway, D.J. (1986) Herbivores in arid and semi-arid rangelands. In: Russell, J.D. and Isbell, R.F. (Eds), *Australian Soils: The Human Impact*, University of Queensland Press, St Lucia, Qld, Australia, pp. 243–71.

Noy-Meir, I. (1973) Desert ecosystems: environment and producers. *Ann. Rev. Ecol. Syst.* **4**, 25–41.

Noy-Meir, I. (1974a) Desert ecosystems: higher trophic levels. *Ann. Rev. Ecol. Syst.* **5**, 195–214.

Noy-Meir, I. (1974b) Stability in arid ecosystems and the effects of man on it. In: *Proceedings First International Congress of Ecology, The Hague*, Centre for Agricultural Documentation, Wageningen, The Netherlands, pp. 220–5.

Noy-Meir, I. (1975) Stability of grazing systems: an application of predator–prey graphs. *J. Ecol.* **63**, 459–81.

Noy-Meir, I. (1980) Structure and function of desert ecosystems. *Isr. J. Bot.* **28**, 1–19.

Noy-Meir, I. (1981) Spatial effects in modelling of arid ecosystems. In: Goodall, D.W. and Perry, R.A. (Eds), *Arid Land Ecosystems*, vol. 2, Cambridge University Press, Cambridge, pp. 411–32.

Noy-Meir, I. (1985) Desert ecosystem structure and function. In: Evenari, M., Noy-Meir, I. and Goodall, D.W. (Eds), *Ecosystems of the World*, vol. 12A: *Hot Deserts and Arid Shrublands*, Elsevier, Amsterdam, pp. 93–104.

Odum, E.P. (1983) *Basic Ecology*, W.B. Saunders, Philadelphia, PA.

Odum, E.P. (1985) Trends expected in stressed ecosystems. *BioScience* **35**, 419–22.

Odum, E.P. and Biever, L.J. (1984) Resource quality, mutualism and energy partitioning in food chains. *Am. Nat.* **124**, 360–76.

Oksanen, L. (1988) Ecosystem organization: mutualism and cybernetics or plain Darwinian struggle for existence. *Am. Nat.* **131**, 424–44.

Olsson, L. (1985) *An Integrated Study of Desertification*. Studies in Geography, No. 13. University of Lund, Sweden.

Otterman, J. (1974) Baring high albedo soils by overgrazing: a hypothesized desertification mechanism. *Science* **186**, 53–533.

Otterman, J. (1981) Satellite and field studies of man's impact on the surface in arid regions. *Tellus* **33**, 68–77.

Otterman, J. and Tucker, C.J. (1985) Satellite measurements of surface albedo and temperature in semi-deserts. *J. Climate Appl. Meteor.* **24**, 228–35.

Perry, R.A. (1968) Australia arid rangelands. *Ann. Arid Zone* **7**, 243–9.

Pickup, G. (1985) The erosion cell—a geomorphic approach to landscape classification in range assessment. *Aust. Rangel. J.* **7**, 114–21.

Pimm, S.L. (1982) *Food Webs*, Chapman & Hall, New York.

Reining, P. (1978) *Handbook on Desertification Indicators*, American Association for the Advancement of Science, Washington, DC.

Rasool, S.I. (1984) On the dynamics of deserts and climate. In: Houghton, J.T. (Ed.), *The Global Climate*, Cambridge University Press, Cambridge.

Ritchie, J.C. and Haynes, C.V. (1987) Holocene vegetation zonation in the eastern Sahara. *Nature* **330**, 645–7.

Ruess, R.W. (1987) The role of large herbivores in nutrient cycling of tropical savannahs. In: Walker, B.H. (Ed.), *Determinants of Tropical Savannas*, ISCU Press, Miami, FL, pp. 67–91.

Sandford, S. (1983) *Management of Pastoral Development in the Third World*, John Wiley & Sons, New York.

Shmida, A., Evenari, M. and Noy-Meir, I. (1986) Host desert ecosystems: an integrated view. In: Evenari, M., Noy-Meir, I. and Goodall, D.W. (Eds), *Ecosystems of the World*, vol. 12B: *Hot Deserts and Arid Shrublands*, Elsevier, Amsterdam, pp. 379–88.

Sinclair, A.R.E. (1975) The resource limitation of trophic levels in tropical grassland ecosystems. *J. Anim. Ecol.* **15**, 497–520.

Sinclair, A.R.E. and Fryxell, J.M. (1985) The Sahel of Africa: ecology of a disaster. *Can. J. Zool.* **63**, 987–94.

Sinclair, A.R.E., Dublin, H. and Borner, M. (1985) Population regulation of the Serengeti Wildebeest: a test of the food hypothesis. *Oecologia* **65**, 266–8.

Skarpe, C. and Bergstrom, R. (1986) Nutrient content and digestibility of forage plants in relation to plant phenology and rainfall in the Kalahari, Botswana. *J. Arid Environs* **11**, 147–64.

Southwood, T.R.E. (1977) Habitat, the template for ecological strategies. *J. Anim. Ecol.* **46**, 337–65.

Spooner, B. and Mann, H.S. (Eds) (1982) *Desertification and Development: Dryland Ecology in Social Perspective*, Academic Press, London.

Starfield, A.M. and Bleloch, A.L. (1986) *Building Models for Conservation and Wildlife Management*, Macmillan, New York.

Tinley, K.L. (1982) Influence of soil moisture balance on ecosystem patterns in Southern Africa. In: Huntley, B.J. and Walker, B.H. (Eds), *Ecology of Tropical Savannahs*, Springer-Verlag, Berlin, pp. 175–91.

UNCOD (1977) *Desertificaton: its Causes and Consequences*, Secretariate of United Nations Conference on Desertification (Ed.) Pergamon Press, Oxford.

United Nations (1980) *Desertification* (Biswas, M.K. and Biswas, A.K. Eds), Pergamon Press, Oxford.

Verstraete, M.M. (1986) Defining desertification: a review. *Climatic Change* **9**, 5–18.

Vukovich, F.M., Toll, D.L. and Murphy, R.E. (1987) Surface temperature and albedo relationships in Senegal derived from NOAA-7 Satellite data. *Remote Sens. Environ.* **22**, 413–21.

Wade, N. (1974) Sahelian drought: no victory for western aid. *Science* **185**, 234–7.

Walker, B.H. (1987) A general model of Savanna structure and function. In: Walker, B.H. (Ed.), *Determinants of Tropical Savannas*, ISCU Press, Miami, FL.

Walker, B.H. and Noy-Meir, I. (1982) Aspects of the stability and resilience of savanna ecosystems. In: Huntley, B.J. and Walker, B.H. (Eds), *Ecology of Tropical Savannas*, Springer-Verlag, Berlin, pp. 556–90.

Warren, A. and Maizels, J.K. (1977) Ecological change and desertification. In: United Nations (Eds), *Desertification*, Pergamon Press, Oxford, pp. 171–260.

Werger, M.J.A. (1986) The Karoo and southern Kalahari. In: Evenari, M., Noy-Meir, I. and Goodall, D.W. (Eds), *Ecosystems of the World*, vol. 12B: *Hot Deserts and Arid Shrublands*, Elsevier, Amsterdam, pp. 283–360.

Westoby, M. (1980) Elements of a theory of vegetation dynamics in arid rangelands. *Isr. J. Bot.* **28**, 169–94.

Wolf, E.C. (1986) Managing rangelands. In: *State of the World*, W.W. Norton, New York, pp. 62–77.

Yodzis, P. (1988) The indeterminancy of ecological interactions as perceived through perturbation experiments. *Ecology* **69**, 508–515.

5 Acid Rain—a Large-scale, Unwanted Experiment in Forest Ecosystems

E.-D. SCHULZE* and B. ULRICH†
**Lehrstuhl Pflanzenökologie, Universität Bayreuth, Postfach 101251, D-858 Bayreuth, Germany*
†Institut für Bodenkunde und Waldernährung, Universität Göttingen, Büsgenweg 2, D-340 Göttingen, Germany

5.1 INTRODUCTION

Since the end of the last century forest ecosystems of central Europe have received large inputs of acidity, nutrients, and toxic substances via dry and wet deposition. Forests appear to have responded to these inputs by altering groundwater relations, soil chemistry, biomass production, species diversity, and by exhibiting forest decline. In order to relate these changes to the effect of 'acid rain', as a general expression for all these anthropogenic inputs, a number of deficiencies in the design of this unwanted 'experiment' have had to be accommodated:

1. The initial states of the ecosystems have had to be reconstructed from sparse historic sources.
2. Since all ecosystems are affected in mid-Europe, there is no untreated control.
3. Both the quantity and the quality of the acid rain input has changed through time—only rough estimates exist of the total deposition including dry deposition.
4. The treatment by acid rain concentrations has been low and within the range of other natural perturbations; therefore, ecosystem responses have been very complex and difficult to unravel. This difficulty has been amplified by the fact that only a few time sequences have been measured, and the variability in space of ecosystems is presently being analyzed instead, and the result is used to interpret the sequence of events in time which led to the observed change.
5. Although any experiment should be repeated, this has not been possible in acid rain research. Acid rain has occurred on a large geographic scale, and covers a time span of more than three generations of researchers.

Ecosystem Experiments. Edited by H.A. Mooney *et al.*

6. At least research on acid rain has tested both a response and a recovery. These have been: irrigation with acidity of unaffected forest land (Tamm, 1989), and liming of acid rain-affected habitats (e.g. Kaupenjohann *et al.*, 1989).

In the following sections we present evidence of how acid rain may have affected forest ecosystems and tree health (Schulze, 1989; Ulrich, 1989), and draw conclusions for future experimental work in ecosystems. We are dealing with areas which exhibit features of forest decline, namely needle loss and needle yellowing, which is associated with various types of nutrient deficiencies and effects on root systems.

5.2 THE INITIAL AND THE PRESENT CHEMICAL STATE OF FOREST SOILS

Before industrialization low-base saturation in podsols in mid-Europe, that had developed in the post-glacial period, was restricted to the A-horizon, while the subsoils had intermediate base saturation. In the 1920s the pH value of B horizons of soils in sheltered areas in north Germany was mainly above 5.0. Even in the 1950s most montane beech and coniferous forests of Germany had a base saturation above 20% and 30% (Ulrich, 1986; Ulrich and Meyer, 1987; Hartmann and Jahn, 1967).

The most convincing evidence that soils have been progressively acidified during the past 20–30 years has been presented by Hallbäcken and Tamm (1986), in which study the soil pH of identical soil profiles was measured in 1927 and 1983. Together with the soil pH measurements of Falkengren-Grerup (1986) and the soil chemical analyses of Ulrich *et al.* (1989), it emerges (Fig. 5.1), that the main change in soil chemistry occurred after 1960. The Ca/Al or Mg/Al ratios in soil solutions decreased in parallel with the change in pH. Presently more than half of the forest soils in northern Germany exhibit a base saturation of exchangeable Ca and Mg below 10% (Ulrich, 1987).

5.3 THE LOAD OF ANTHROPOGENIC ACIDS

The total accumulated acid load deposited into forest ecosystems since the beginning of industrialization in the northern and central part of West Germany has been estimated on the basis of emission and deposition data to vary between 6 and >34 mol acid equivalents m^{-2} ground area. However, in order to evaluate the effect of this proton load on ecosystems it is important to recognize that the chemistry of acid precipitation has changed with time

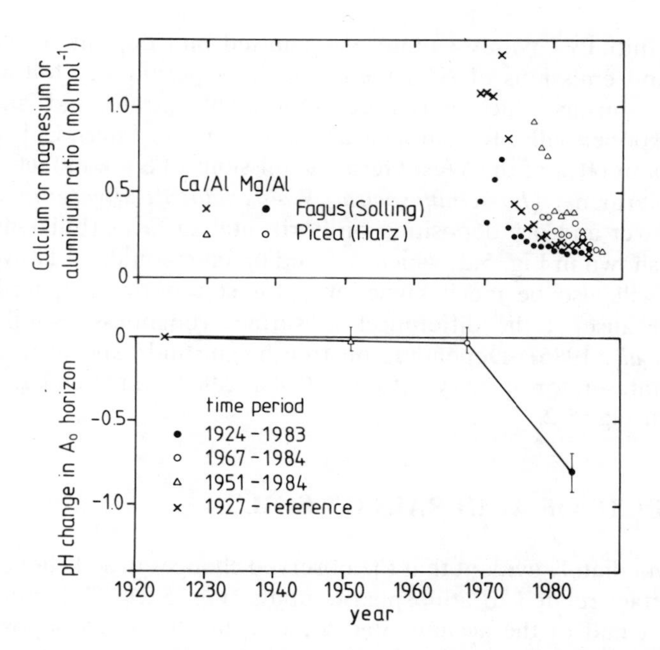

Figure 5.1 Yearly average ratios of Ca and Mg to Al measured since 1968 in the Harz and the Solling mountains, north Germany (top), and the pH in the A_0 horizon from 1924 until 1985 (bottom). Error bars indicate standard errors ($n = 41$ for the period 1924–83, $n = 21$ for period 1967–84, $n = 38$ for period 1951–84 (from Schulze, 1989).

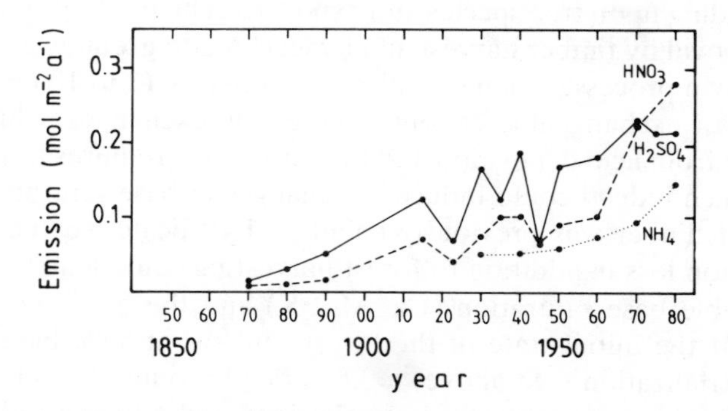

Figure 5.2 The change of emission and deposition of sulfate, nitrate, and ammonium since the year 1870. The rates of sulfate and nitrate were calculated on the basis of emissions by Ulrich and Meyer (1987) as equivalents from emission data on an area basis of Germany. The ammonium deposition data are for the Netherlands (Boxman *et al.*, 1988).

(Fig. 5.2). Initially, mainly sulfate was emitted and deposited, while in a second period emissions of NO_x increased. This period was followed by a third phase during which aerial depositions of nitrate and ammonium increased exponentially. Recent measurements across western Europe reveal that only about 60% of the West German emissions of S are in fact deposited in West Germany (*Forschungsbeirat Waldschäden/Luftverunreinigungen*, 1989). The wet and dry deposition on horizontal surfaces (bulk sampler) is lower than shown in Fig. 5.2, which is based on emission data. However, dry deposition will also be much higher in a forest canopy compared to bulk samplers because of the differences in surface roughness (Schulze, 1989; Lindberg *et al.*, 1986). Depending on tree height and exposure, the actual deposition rates in forests vary between 20% and 90% of the emission density presented in Fig. 5.2.

5.4 EFFECTS OF ACID RAIN ON SOILS

It is not immediately evident that the observed changes in soil chemistry (Fig. 5.1) are in fact related to atmospheric inputs (Fig. 5.2). The emission data indicate that half of the accumulated total acidity has been deposited after 1950, and one-third after 1970.

A very general estimate of a stand proton balance (Ulrich, 1989) indicates that during a 100-year rotation period a forest will accumulate 5–7 mol m^{-2} cation equivalents in its biomass and timber, which is harvested and removed from that site. The system could eventually lose up to 10 mol cation equivalents m^{-2} ground area from natural leaching of nitrate, and it has received a cumulative acid deposition between 6 and 34 mol H^+ m^{-2} from the atmosphere depending upon tree species and exposure. The total loss of 10–40 mol H^+ m^{-2} (removal by timber harvest and by leaching to groundwater) must be balanced by two processes, namely silicate weathering (2 to 10 mol H^+ m^{-2}) and leaching of exchangeable cations from cation exchanges. This indicates that the input from acid deposition and timber harvest requires large amounts of cations which indeed could induce the changes in base saturation of soils. Harvesting of timber was previously balanced by silicate weathering. However, any cation loss in addition to forest harvesting could lead to a decrease in exchangeable base saturation (Ca, Mg, K) and the observed changes in forest soils. If the initial state of the forest soil was at 50% base saturation before industrialization (see above) a 0.5 m deep profile of sandy soil would store about 2 mol m^{-2} of exchangeable basicity and a loamy soil between 6 and 30 mol m^{-2}. If the cation requirement for timber growth and soil leaching removed 10–40 mol cation equivalents m^{-2} then low base saturation may be reached in forest soils depending on the rate of weathering.

The acid/base reactions of soils depend on the chemical nature of the

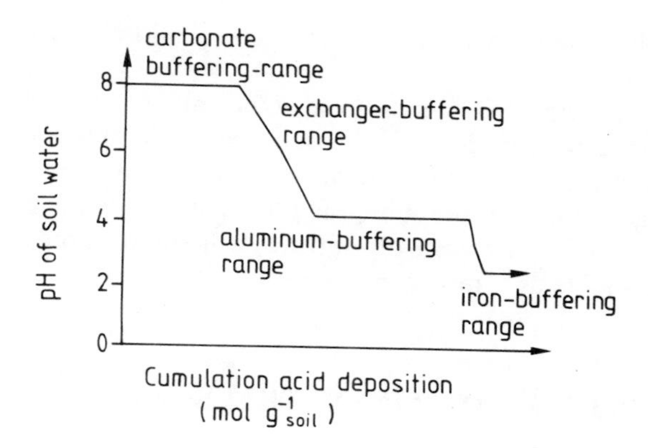

Figure 5.3 Schematic presentation of the change in soil pH during cumulative exposure by acid deposition. See text (after Prenzel, 1985).

substrate, the incoming acid, and the pH in the soil solution. Only if the pK of the acid is below the pH of the soil solution it will contribute to the acidification process. In this case the acid reacts with the soil matrix, which buffers the soil solution against changes in pH. In a humid climate, soils exhibit a characteristic pH response to an accumulated deposition of strong acids (Fig. 5.3; Ulrich, 1981; Van Breemen *et al.*, 1983). Between pH 8.6 and 6.2, buffering is primarily by calcium carbonate, which becomes dissolved during the buffering process. Once this reservoir becomes exhausted, soils enter an 'exchange-buffering range', which extends to pH 4.2 and in which mainly clay minerals progressively exchange cations for hydrogen ions which release aluminum ions from the clay lattice. Both cation exchange capacity and pH decrease with continued leaching of base cations, and exchangeable Al ions finally become the dominant reactive cation species. In case of an input of sulfuric acid, this acid may be immobilized by forming aluminum alone, such as nitrate or organic acids (which may also occur naturally in soil) can continue to promote the acidification process below pH 4.2 and allow Al ions to enter the soil solution. Thus, sulfates are accumulated in the soil profile during the acidification process at higher base saturation, but may be mobilized again when soils acidify below pH 4.2 (Ulrich, 1987). This has been demonstrated by input/output budgets (Ulrich, 1989, *Forschungsbeirat Waldschäden*, 1989).

In a soil–chemical model Kaupenjohann (1989) demonstrated that sulfate deposition alone will not acidify soils below pH 4.2 because of binding with Al. However, as soon as other acids are leached, such as nitric acid, the acidification process is strongly enhanced. Since the uptake of nitrate by roots reverses the acidification which took place during nitrification it is important to understand why nitrate is not used by the plant cover. If plants consumed nitrate by uptake, the increase in acid strength would be terminated by this plant process.

It becomes clear that the 'ecosystem experiment' of acid rain is quite complicated, since we are dealing not only with soils but also with a plant cover, and because sulfate and nitrate have been deposited in a changing ratio (Fig. 5.2). In addition, ammonium was added in changing amounts to the total deposition which, in contrast to nitrate, will enhance the acidification of soils if used by plants as a nitrogen source.

5.5 EFFECTS OF ACID RAIN ON TREES

If cation deficiency in the soil were the only factor acting on plant performance, then it would be expected that plant growth would respond to this deficiency, and the nutrient ratios should remain constant due to a regulated and balanced nutrient uptake by plant roots (Ingestad, 1987). This, however, is not the case under conditions of acid rain. The process of plant adjustment appears to be disturbed since nutrient imbalances are one of the many observed symptoms of forest decline (Oren and Schulze, 1989; Schulze, 1989).

5.5.1 EFFECTS OF SOIL ACIDIFICATION ON ROOT PERFORMANCE

The change in the Mg/Al and Ca/Al ratio in the soil solution during soil acidification (Fig. 5.1) affects root growth (Rost-Siebert, 1985). The primary effect appears to be a substitution of Ca and Mg by Al in the secondary cell wall, which affects its extension growth (Jorns, 1988). As a result there is a correlation between the number of root tips and the Ca/Al ratio in chemical parameters (Meyer *et al.*, 1986). In addition to the interaction with Al, root growth is also affected by a changing NH_4/cation ratio (Bertiller, 1986). An increase of ammonium in the soil solution of the humus layer has been observed in acidifying soils (Fig. 5.4; Schneider *et al.*, 1989) which is probably a result of the recent increase of deposition of ammonium from the atmosphere as well as from a reduced microbial activity in acid soils.

The Mg/Al and Mg/NH_4 ratios affect not only root growth, but also the uptake of cations. Ca and Mg are reduced at the presence of Al and NH_4 (Schröder *et al.*, 1988; Jacob, 1955). In addition, nitrate uptake is inhibited in the presence of NH_4 (Roeloffs *et al.*, 1985, 1988). The ratios of stable isotopes of $^{15}N/^{14}N$ indicate that, under conditions of low base saturation, conifers use ammonium rather than nitrate from the soil solution (Schulze and Gebauer, 1989).

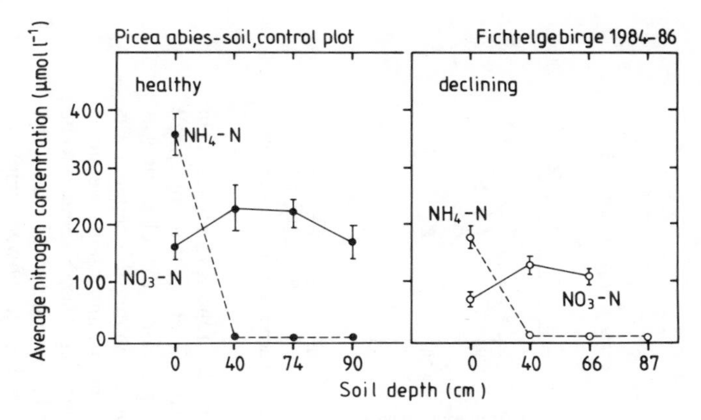

Figure 5.4 Concentrations of ammonium and nitrate in soil profiles of a healthy and declining spruce stand in West Germany (from Schneider *et al.*, 1989).

The results on ion uptake of roots indicate that the leaching of nitrate from soils, which contributes strongly to soil acidification, results in essence from a plant-regulated physiological process, namely the selective uptake of ammonium by tree roots. Soil acidification, that initially started by sulfate leaching, had a feedback on the form of nitrogen used by the plant cover. An increasing NH_4/NO_3 ratio in the soil solution resulted in preferential uptake of ammonium. This again had a positive feedback on the leaching of nitrate and the acidification process (Schulze, 1989).

Low supply of cations alone cannot account for the observed expression of nutrient deficiency symptoms during forest decline, because if all nutrients were equally deficient for optimal growth, plants would adjust their growth within a certain range to match the supply of nutrients and show no deficiency symptom (Ingestad, 1987). In order to understand the interaction between growth and cation nutrition Oren *et al.* (1988) have applied the concept of Timmer and Armstrong (1987) and demonstrated that newly formed needles display nitrogen deficiency symptoms rather than optimal supply (Fig. 5.5). At the same time, Mg had been incorporated to new needles to such an extent that even luxury levels were reached! This indicates that these trees received N far below their optimum level for growth, even though nitrate was leaching out of the soil profile, again pointing to a selective uptake of ammonium.

Only 30–50% of the Mg incorporated in new growth originates from concurrent root uptake (Oren *et al.*, 1988). The larger proportion is reallocated from available pools in the existing tree biomass (Fig. 5.6), especially from the 1–2-year-old needles. When new twigs and needles start to grow, reallocation will be proportional to the supply of N and independent of the initial Mg concentration in the old needles. When initial Mg content of

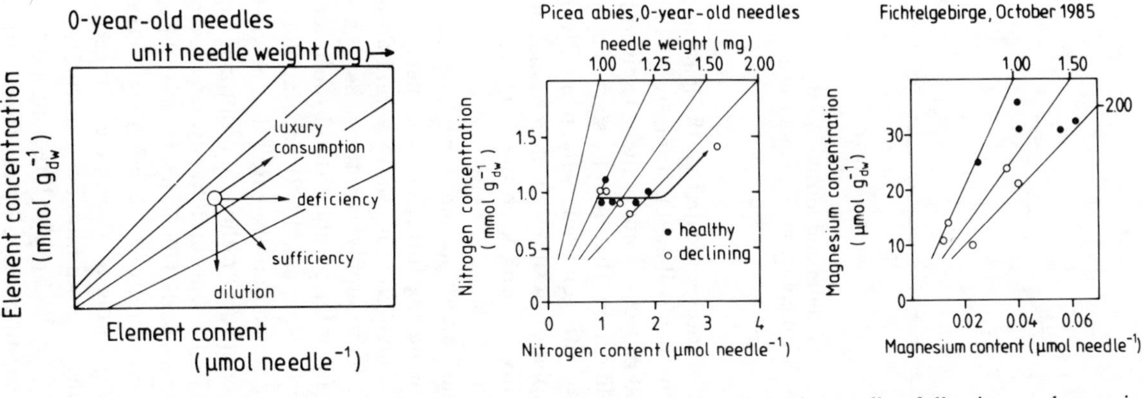

Figure 5.5 Diagnosis of nutrient limitation based on the responses of single needles following a change in nutrient supply (left). If element concentration and element content both increase, this element is consumed luxuriously. If the element concentration remains constant but element content increases, this element was present at a deficiency level. If the element concentration decreases at constant element content, this element was diluted. Relations between needle N (middle and Mg (right) concentrations, and content, and needle weight in current year needles from 10 plots of healthy and declining *Picea abies*. Based on this diagnosis, N seemed to be deficient except in one plot, in which Mg stress prevented further needle growth and N accumulated luxuriously. Mg seemed to be in excess in current-year needles although mature needles at the declining site showed symptoms of Mg deficiency (from Schulze, 1989).

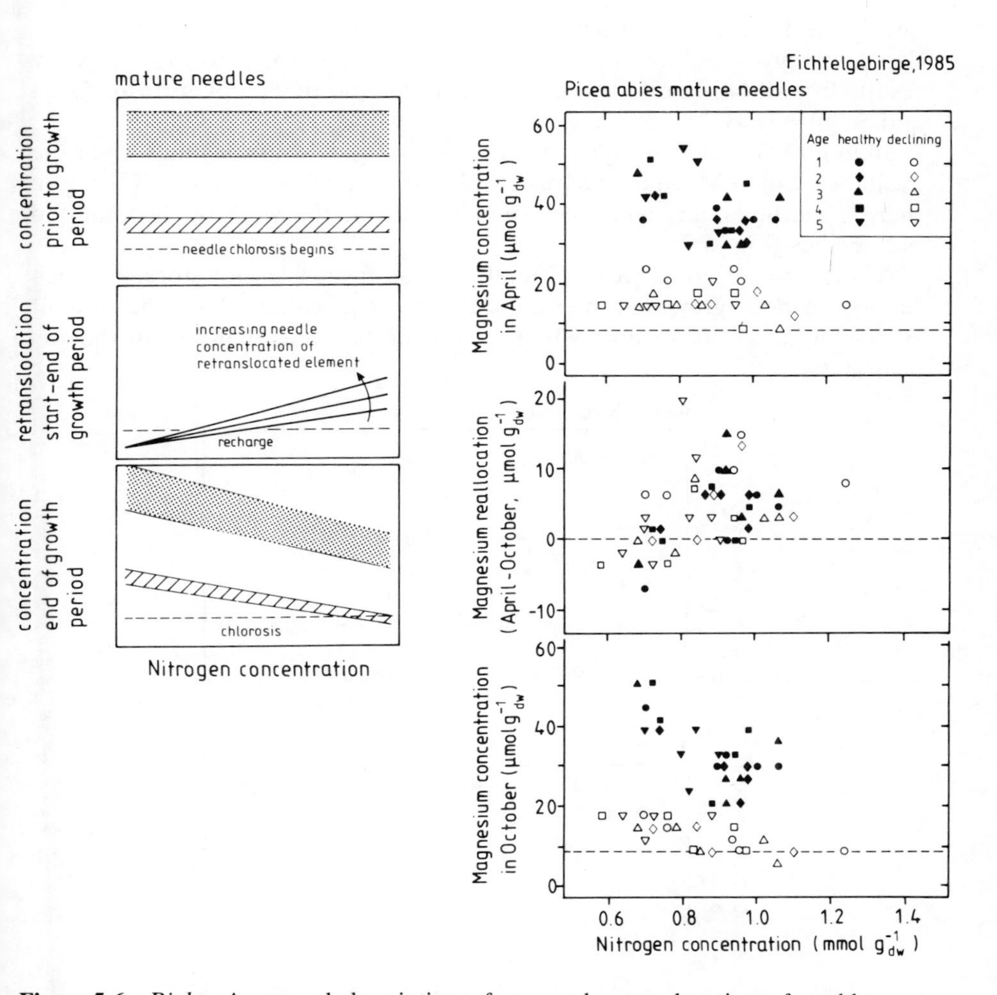

Figure 5.6 *Right*: A general description of seasonal retranslocation of a phloem-mobile element from mature needles in relation to the N concentration in those needles. Initial concentrations at the start of the growing season (top) were high (shaded area) for needles that were able to accumulate large amounts of that element before the growth phase began. They were low for needles that were not able to accumulate that element (hatched area). With the beginning of canopy growth (middle), the mobile element is retranslocated out of mature needles in relation to N concentration. More N promotes growth and increases demand for other elements. As a result, the element concentration at the end of the growth phase is inversely related to the N concentration (bottom), and this may result in needle nutrient deficiency and yellowing. *Left*: Measurements are on trees of *Picea abies* of the processes described above (from Schulze, 1989).

needles is low, mobilization of Mg from old needles, in response to N-dependent growth of new needles, will lead to yellowing. Thus Mg deficiency results from a N/Mg interaction during the period of canopy growth (Oren and Schulze, 1989).

Stem growth tends to follow needle growth, and occurs at times when the available pool of Mg in the biomass is likely to be almost depleted. Therefore there is a strong relation between stem growth and the Mg/N or Ca/N ratio in old needles (Fig. 5.7).

After termination of growth, uptake of cations is used for storage, which supports canopy growth in the next season by reallocation. Also, the amount of re-charge will determine whether a tissue becomes deficient in the wet season.

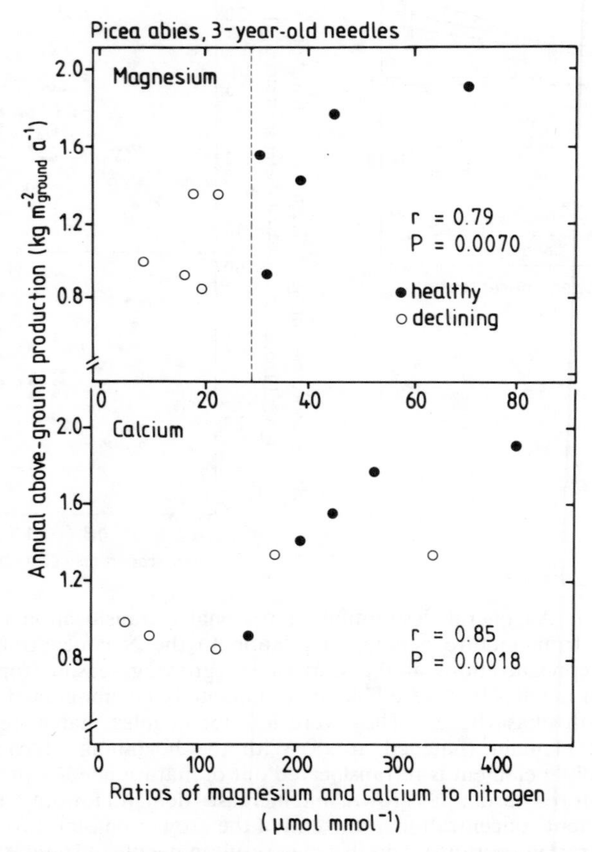

Figure 5.7 Annual above-ground production on the plots of healthy and declining *Picea abies* as related to the ratio of Mg/N and Ca/N in 3-year-old needles. The dashed line indicates Mg deficiency according to the laboratory experiments of Ingestad (1959) (from Oren *et al.*, 1988).

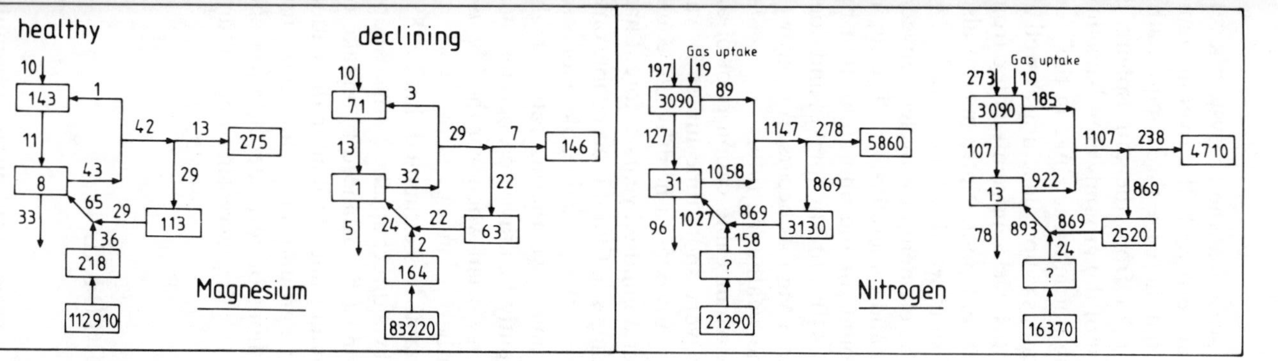

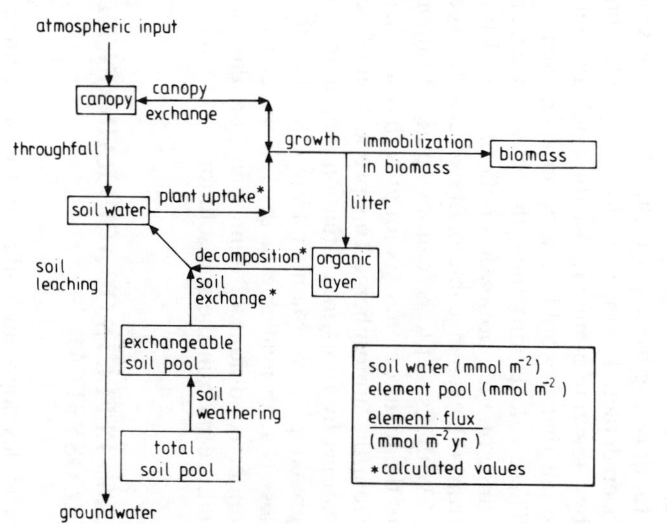

Figure 5.8 Fluxes and N (right) pools in a healthy and declining forest site. The data are based on five plots at each site in the Fichtelgebirge, West Germany. The boxes indicate element pools, the arrows indicate element fluxes (from Schulze, 1989).

5.5.2 NUTRIENT CYCLING

It is important to integrate these individual processes at the ecosystem level. Only the balance of the ecosystem internal cycle reveals the overall significance of various processes. Figure 5.8 shows that there are three major paths of element flow (Horn *et al.*, 1989). The input from the atmosphere is connected via canopy throughfall to the output of the groundwater. This flux is connected to the plant internal cycle during nutrient exchange in the tree canopy and in the soil water. Plant uptake supplies canopy leaching as well as growth, litter fall, and mineralization. Into this cycle nutrients enter from weathering of the exchangeable and total nutrient pool. Nutrients are also immobilized from the ion cycle by growth of stem wood for harvest.

With respect to Mg, a large proportion of the element flow cycles between soil solution, plant uptake, canopy litter fall, and mineralization. The input from the atmosphere is large enough to account for the immobilization by growth. Canopy leaching of Mg accounts for 2–10% of Mg uptake, and does not seem to influence the overall flux. The main Mg loss occurs from growth of timber and soil leaching, which exceeds the weathering rate.

According to this estimated balance, the exchangeable cation pool of an apparently 'healthy' site would be used in a very short time unless it was replaced by weathering of the total soil pool. However, a healthy site could still draw Mg from a large 'total pool' for several hundred years before it had been depleted to the same level as at a 'declining' site (but it is accessible only by a low rate of mineral weathering). In contrast, the exchangeable pool is so small that it would be used in a few years, unless weathering rate is high. These calculations show that even an apparently healthy site operates at a fringe of a temporary Mg deficiency, which may occur whenever the Mg use from the exchangeable pool exceeds the weathering rate.

With respect to N, canopy uptake is important. It is estimated that 8–18% of the requirement of N for growth is supplied via direct uptake by the canopy from the atmosphere. This proportion of N must be balanced by uptake of cations by roots in a situation where ammonium and Al hinder the cation uptake process. Mainly nitrate is leached to the groundwater, since roots use less N than available in the soil under conditions of aerial fertilization and under conditions of ammonium uptake due to the increasing ammonium/nitrate ratio in the soil solution.

5.6 EFFECTS OF ACID RAIN ON THE TOTAL ECOSYSTEM

The development leading to forest decline is characterized by the accumulation of acidity deposited in the soil from the very beginning of soil formation,

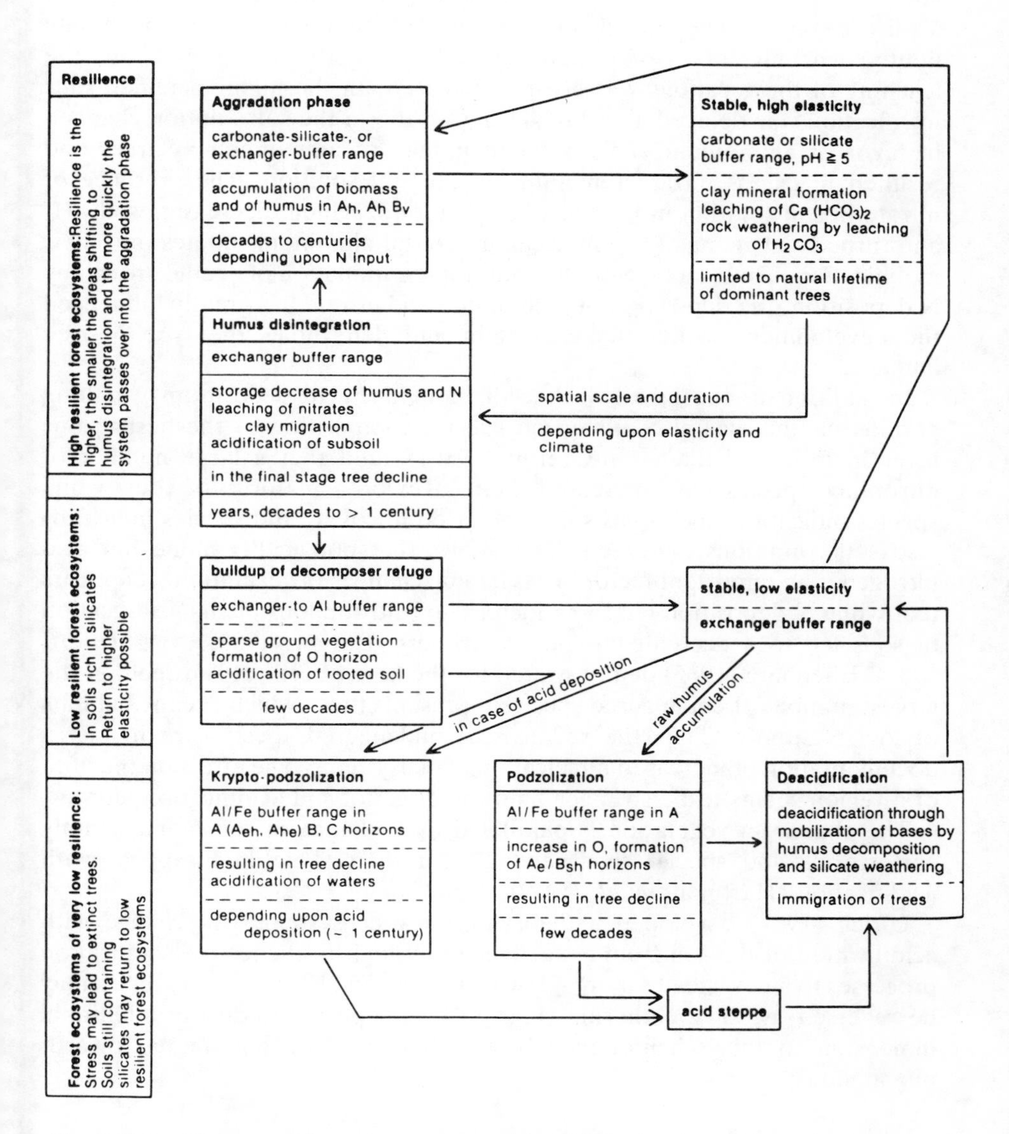

Figure 5.9 Changes of forest ecosystems with acidification in relation to stability and resilience (from Ulrich, 1987).

and by increasing deposition rates, first of sulfate and nitrate (from 1900 on), later also of ammonium (from 1960 on). Up to 1960 the accumulation of acidity in soils occurred mostly in the form of aluminum sulfate, i.e. with only limited leaching losses of exchangeable Ca and Mg. From 1960 on, the leaching of these cations was greatly increased. In the highly acidified soils nitrification was delayed and the NH_4/NO_3 ratio in the soil solution changed in favor of ammonium. This initiated uptake of ammonium by trees and resulted in excess nitrate remaining in the soil solution. The presence of nitrate in soil solution in turn increased cation leaching, decreasing soil pH, and turned soil chemistry from accumulation of aluminum sulfates into their mobilization. Thus, along with the continuing stimulation of needle growth by N deposition, the deposition of ammonium and nitrate has greatly influenced the development within the ecosystem and determines the type of tree damage.

In addition to the forest decline phenomena, acid deposition over long periods of time has also influenced species composition of the herbaceous layer in forests. Falkengren-Grerup (1986) found that a large number of nitrophilic species have increased their cover on a majority of sites, while species indicating calcareous soils decreased in cover, and species indicating acid soil conditions increased. This shows that not acidity alone but also nitrogen is an important factor in explaining changes in vegetation. It appears that while N deposition led to an increase of nitrophilous plants acidification of soils led to a decrease in species richness (Bürger, 1988; Rost-Siebert, 1986). Ellenberg (1985) demonstrated for the flora of central Europe that the largest number of endangered species represents those which occupy habitats of low N supply. Thus the red lists of endangered species are in fact a documentation of increased nitrification of ecosystems. The effect on the flora of a region seems to be stronger than the effect of soil acidification alone.

The change in vegetation is paralleled by a change in soil fauna; mainly earthworms and species that need Ca for their skeleton disappear with acidification (Hartmann *et al.*, 1989).

Ulrich (1987) developed a theory on how ecosystems will change with acidification of the soil from a steady state of high resilience by acidification processes to a new steady state of low resilience, and how these processes may be reversed by soil weathering (Fig. 5.9). N-deposition adds an additional dimension to this scheme, namely a change in flora and fauna towards nitrophilous species.

5.7 CONCLUSIONS: WHAT CAN WE LEARN FROM ACID RAIN WITH RESPECT TO ECOSYSTEM EXPERIMENTS?

The processes involved during forest decline by air pollution are summarized in Fig. 5.10. It becomes obvious that the unwanted ecosystem experiment of

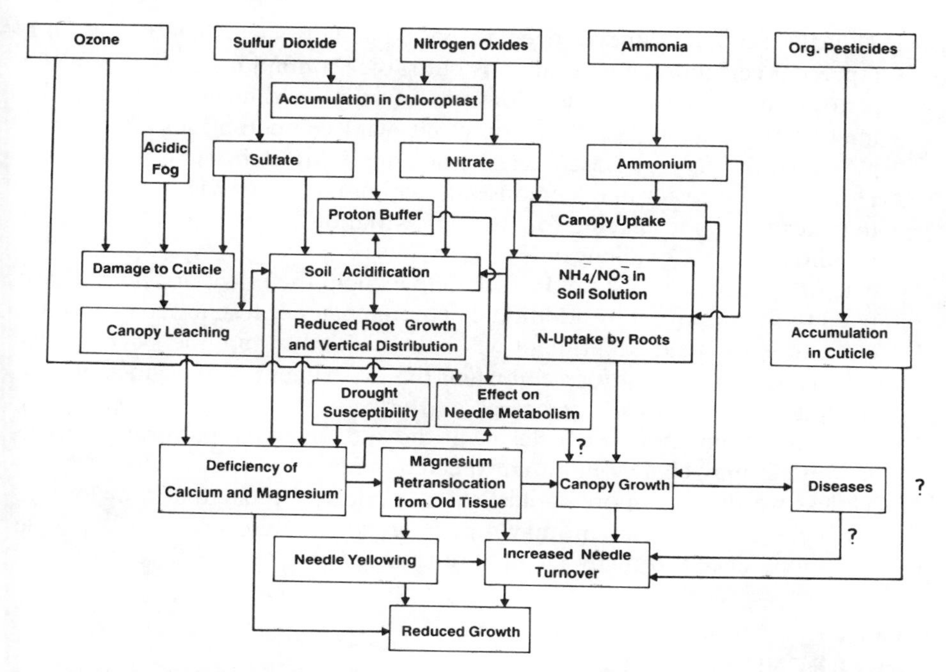

Figure 5.10 Schematic presentation of processes that led to the forest decline symptoms following acid rain. The boxes indicate the air pollutants and their effect in the soil and in the plant. The arrows indicate the main effects which were identified. The scheme is not complete with respect to all interactions (from Schulze, 1989).

acid rain has not initiated only a single process, but rather a whole network of effects. Besides acid rain, gaseous air pollutants have also affected forest ecosystems. Additionally, biotic factors became secondary damaging agents. The results are more complicated than was anticipated. The effect on the total system cannot be understood by simple treatment/response reaction analysis. For future ecosystem experiments we may learn from this event:

1. The results are based on a 100-year treatment. However, they are still observations, and do not represent cause and effect in an experimental sense. All factors which correlate with each other could in turn be related to another set of factors. The interaction of ammonium and nitrate may be such an example. Thus all statistical relationships require additional experimental perturbations in order to prove cause and effect.
2. Correlation analysis is difficult to apply whenever an organism regulates a process to be constant. For example, the nitrogen concentration in needles remained constant, since all nitrogen was used for growth. Thus, N concentration of needles may not correlate with N deposition, because it is a plant-regulated compartment.

3. The observational studies need to be expanded, such that they do not represent certain specific situations but rather continuous gradients of the factors under investigation. However, it may be difficult to define the important gradients because of internal plant compensations. For example, a study may be based on a gradient of Mg deficiency in needles. However, the result may demonstrate an interaction of Mg with N. Since the sites were not selected for studying variations in N, a new study must be initiated on a gradient in N.
4. The observations in the field which help explain the effect of acid rain still need to be supported by additional experiments, because it becomes quite clear that any field experiment will have numerous and unexpected side-effects, which are not foreseen when the experiment starts. Since we are dealing with a network of causes and effects, the interpretation of any experiment may not be conclusive at the end, since the pathway of cause and effect may change in an unexpected manner.
5. Unless we develop more sophisticated models of ecosystem functioning, any ecosystem experiment may remain an *ad-hoc* description of a specific situation, since too many factors are changed at the same time.

REFERENCES

Bertiller, M.B. (1986) Wachstums- and Schädigungsdynamik von Fichtenwurzeln als Funktion der bodenchemischen Bedingungen. Ber. *Forschungszentrum Waldökosysteme/Waldsterben Göttingen*, Series A, Volume 21.

Boxman, D., Dijk van, H. and Roelofs, J. (1988) Critical loads for nitrogen, with special emphasis on ammonium. In: Nilsson, I. and Grennfelt, P. (Eds), *Critical Loads for Sulfur and Nitrogen, Miljörapport* **15**, pp. 295–322.

Breemen, N. van, Mulder, J. and Driscoll, C.T. (1983) Acidification and alkalinization of soils. *Plant Soil* **75**, 282–308.

Bürger, R. (1988) Veränderungen der Bodenvegetation in Wald- und Forstgesellschaften des mittleren und südlichen Schwarzwaldes. KfK-PEF 52.

Ellenberg, H., Jr (1985) Veränderungen der Flora Mitteleuropas unter dem Einfluß von Düngung und Immissionen. *Schweiz. Z. Forstwesen* **136**, 19–39.

Falkengren-Grerup, U. (1986) Soil acidification and vegetation changes in deciduous forest in southern Sweden. *Oecologia* **70**, 339–47.

Forschungsbeirat Waldschäden/Luftverunreinigungen (1989) 3. Bericht, Bundesministerium für Forschung und Technologie, Bonn.

Hallbäcken, L. and Tamm, C.O. (1986) Changes in soil acidity from 1927 to 1982–1984 in a forest area in south-west Sweden. *Scand. J. For. Res.* **1**, 219–32.

Hartmann, F.K. and Jahn, G. (1967) *Waldgesellschaften des mitteleuropäischen Gebirgsraums nördlich der Alpen*, Gustav Fischer Verlag, Stuttgart.

Hartmann, P., Scheitler, M. and Fischer, R. (1989) Soil fauna comparisons in healthy and declining Norway spruce stands. *Ecol. Stud.* **77**, 137–50.

Horn, R., Schulze, E.-D. and Hantschel, R. (1989) Nutrient balance and element cycling in healthy and declining Norway spruce stands. *Ecol. Stud.* **77**, 444–58.

Ingestad, T. (1959) Studies on the nutrition of forest tree seedlings. II. Mineral nutrition of spruce. *Physiol. Plant.* **12**, 568–93.

Ingestad, T. (1987) New concepts of soil fertility and plant nutrition as illustrated by research on forest trees and stands. *Geoderma* **40**, 237–52.

Jacob, A. (1955) *Magnesia: der fünfte Pflanzennährstoff*, Ferdinand Enke Verlag, Stuttgart.

Jorns, A.C. (1988) Aluminiumtoxidität bei Sämlingen der Fichte (*Picea abies* (L.) Karst) in Nährlösungskultur. *Ber. Forschungszentrum Waldökosysteme/ Waldsterben, Series A*, Vol. 42.

Kaupenjohann, M. (1989) Effects of acid rain on soil chemistry and nutrient availability. *Ecol. Stud.* **77**, 296–340.

Kaupenjohann, M., Zech, W., Hantschel, R., Horn, R. and Schneider, B.U. (1989) Effects of fertilization. *Ecol. Stud.* **77**, 418–24.

Lindberg, S.E., Lovett, G.M., Richter, D.D. and Johnson, D.W. (1986) Atmospheric deposition and canopy interactions of major ions in a forest. *Science* **231**, 93–192.

Meyer, J., Schneider, B.U., Werk, K.S., Oren, R. and Schulze, E.-D. (1986) Performance of two *Picea abies* (L.) Karst. stands at different stages of decline. V. Root tip and ectomycorrhiza development and their relations to above ground and soil nutrients. *Oecologia* **77**, 7–13.

Oren, R. and Schulze, E.-D. (1989) Nutritional disharmony and forest decline: a conceptual model. *Ecol. Stud.* **77**, 425–43.

Oren, R., Schulze, E.-D., Werk, K.S. and Meyer, J. (1988) Performance of two *Picea abies* (L.) Karst. stands at different stages of decline. VII. Nutrient relations and growth. *Oecologia* **77**, 163–73.

Prenzel, J. (1985) Verlauf und Ursachen der Bodenversauerung. *Z. Deutsch. Geol. Ges.* **136**, 293–302.

Roelofs, J.G.M., Kempers, A.J., Houdijk, A.L.F.M. and Jansen, J. (1985) The effect of airborne ammonium sulfate on *Pinus nigra* var maritima in the Netherlands. *Plant Soil* **84**, 45–56.

Roelofs, J.G.M., Boxman, A.W., Dijk, H.F.G. van and Houdijk, A.L.F.M. (1988) Nutrient fluxes in canopies and roots of coniferous trees as affected by nitrogen-enriched air pollution. In: Bervaes, J., Mathy, P. and Evers, P. (Eds), *Relationships Between Above and Below Ground Influences of Air Pollutants on Forest Trees*, Commission of the European Communities, Brussels.

Rost-Siebert, K. (1985) Untersuchungen zur H- und Al-Ionen Toxidität an Keimpflanzen von Fichte (*Picea abies* Karst.) und Buche (*Fagus sylvatica* L.) in Lösungskultur. *Ber. Forschungszentrum Waldökosysteme/Waldsterben Göttingen*, Vol. 12.

Rost-Siebert, K. (1986) Feststellung von Veränderungen der Bodenvegetation und im chemischen Oberbodenzustand während der letzten Jahrzehnte. *UBA Texte* **18**, 246–56.

Schneider, B.U., Meyer, J., Schulze, E.-D. and Zech, W. (1989) Root and mycorrhizal development in healthy and declining Norway spruce stands. *Ecol. Stud.* **77**, 370–91.

Schröder, W.H., Bauch, J. and Endeward, R. (1988) Microbeam analysis of Ca exchange and uptake in the fine root of spruce: influence of pH and aluminum. *Trees* **2**, 96–103.

Schulze, E.-D. (1989) Air pollution and forest decline in a spruce (*Picea abies*) forest. *Science* **244**, 776–83.

Schulze, E.-D. and Gebauer, G. (1989) Aufnahme, Abgabe und Umsatz von Stickoxiden, NH_4 und nitrate bei Waldbäumen, insbesondere der Fichte. *GSF Bericht* **6**(89), 119–34.

Schulze, E.-D., Oren, R. and Lange, O.L. (1989) Processes leading to forest decline: a synthesis. *Ecol. Stud.* **77**, 459–68.

Tamm, C.O. (1989) Comparative and experimental approaches to the study of acid deposition effects on soils as substrate for forest growth. *Ambio* **18**, 184–91.

Timmer, V.R. and Armstrong, G. (1987) Diagnosing nutritional status of containerized tree seedlings: comparative plant analyses. *Soil Sci. Soc. Am. J.* **51**, 1082–6.

Ulrich, B. (1981) Ökologische Gruppierung von Böden nach ihrem chemischen Bodenzustand. *Z. Pflanzenernährung Bodenkunde* **144**, 289–305.

Ulrich, B. (1986) Natural and anthropogenic components of soil acidification. *Z. Pflanzenernährung Bodenkunde* **149**, 702–17.

Ulrich, B. (1987) Stability, elasticity, and resilience of terrestrial ecosystems with respect to matter balance. *Ecol. Stud.* **61**, 11–49.

Ulrich, B. (1989) Effects of acidic precipitation on forest ecosystems in Europe. *Adv. Environ. Sci.* **2**, 189–272.

Ulrich, B. and Meyer, H. (1987) Chemischer Zustand der Waldböden Deutschlands zwischen 1920 und 1960, Ursachen und Tendenzen seiner Veränderung. *Ber. Forschungszentrums Waldökosysteme/Waldsterben, Series B* **6**, 1–133.

Ulrich, B., Meyer, H., Jänich, K. and Büttner, G. (1989) Basenverluste in den Böden von Hainsimsen-Buchenwäldern in Südniedersachsen zwischen 1954 und 1986. *Forst und Holz* **44**, 251–3.

6 What Can We Learn from Uncontrolled Ecosystem Experiments with Inorganic and Organic Micropollution?

R. HERRMANN
Lehrstuhl für Hydrologie, Universität Bayreuth, D-8580 Bayreuth, Postfach 101251, Germany

6.1 INTRODUCTION

The ever-increasing amount of micropollutants emitted from agriculture, households, and industry into our environments may endanger the functioning of ecosystems. The evaluation of the ecotoxicological impact of the micropollutants calls for an assessment of the combined accumulation, advection, diffusion, and reaction processes that determine the exposure of the biota to the micropollutants within the ecosystem.

My aim is to point out the possibilities and limits to predicting the exposure concentrations that might realistically exist within individual compartments of a terrestrial ecosystem. I concentrate my attention on the compartmental distribution of trace pollutants as a result of more or less long-term and background pollution.

6.2 FATE OF MICROPOLLUTANTS AND THE RESULTING EXPOSURE OF BIOTA

The task of ecotoxicology, namely to assess the risk for an ecosystem exposed to micropollutants, can be looked at from two points of view:

1. The evaluation of the impact of a single micropollutant or a group of different micropollutants upon the functioning of an ecosystem.
2. The assessment of the influence of the environment upon the distribution and fate within an ecosystem (Burns, 1983; Lassiter *et al.*, 1979; Mackay and Paterson, 1982; Stumm *et al.*, 1983).

Ecosystem Experiments. Edited by H.A. Mooney *et al.*

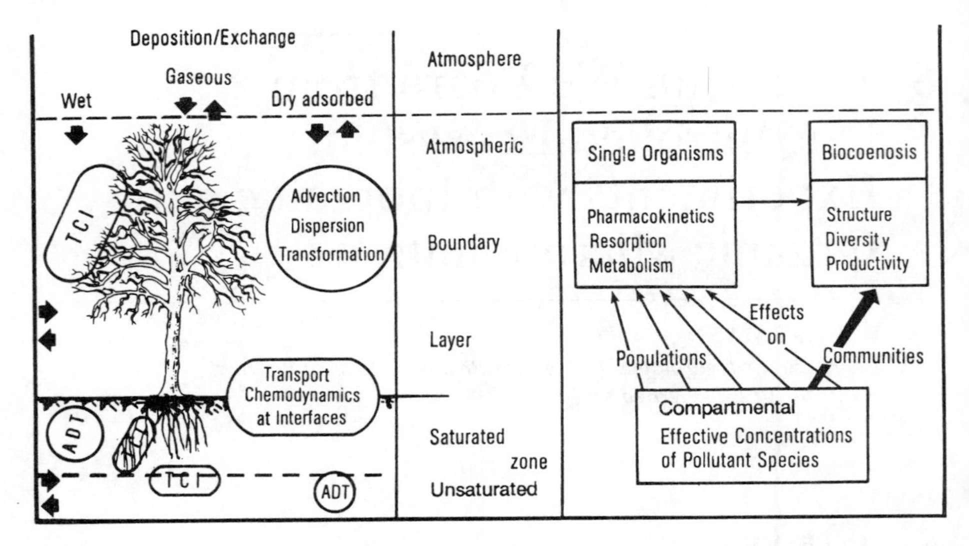

Figure 6.1 Fate of micropollutants within the ecosystem and the resulting exposure of biota (modified from Stumm *et al.*, 1983); TCI = transport and chemodynamics at interfaces; ADT = advection, dispersion, and transformation.

The relationship of the transfer and transformation processes within a terrestrial ecosystem and the resulting exposure of the organisms to the micropollutants is shown in Fig. 6.1. The predominant mechanisms by which micropollutants reach the ecosystem are wet and gaseous deposition and the deposition of micropollutants adsorbed on particles. Within the ecosystem the fate of the micropollutants is governed by various transfer and trans-formation processes, some of which are very difficult to evaluate in the field. Whereas the dissolved or gaseous transport can be determined by measuring fluid velocities and the concentrations, suspended particles may also adsorb the micropollutants. Thus there exist two- or three-phase transfer systems: in the soil, as well as in the aquatic microenvironments of droplets and water-films on leaves or within the soils, the micropollutants may be found in the gaseous phase, dissolved in water or adsorbed onto solid particles. Transport processes will be further complicated by molecular diffusion or volatilization. These and sorption processes are also of great importance for transfer of pollutants across the various air/water, air/solid and water/solid interfaces.

Various transformation processes such as photolysis, hydrolysis, reaction with other chemicals and microbial metabolism make up the chemodynamics of trace pollutants within the ecosystem.

At the solid interfaces we can also observe catalysed chemodynamic processes such as photo-oxidation of organic micropollutants.

Biosorption, bioaccumulation, or leaching are the cause of transport of trace pollutants within the food web. The transfer within the food web, however, is highly variable and depends on direct biosorption from water, dietary exposure, metabolism, and depuration.

Even more than in aquatic environments, studies in terrestrial ecosystems have mainly concentrated on describing the effects of numerous pollutants on various organisms; they have, however, neglected to a great extent to investigate their fates.

6.3 ENVIRONMENTAL ANALYSIS OF MICROPOLLUTANTS

In order to evaluate the direct toxicity of a single micropollutant or a group of micropollutants to individual populations, or the ecotoxicological effects of these pollutants to whole communities, we must be able to trace the fate of these micropollutants through the different compartments of an ecosystem. Further, we must be able to describe the pollutants' degradation, distribution, and bioaccumulation. This information should then be related to direct chronic or acute toxicity, mutagenicity, or carcinogenicity and finally to socio-biological effects.

For this interpretation we need a chemical analysis that is able to distinguish pollutants collectively (e.g. the concentration of all polycyclic aromatic hydrocarbons together) or individually (e.g. the concentration of a single chemical species such as free aluminium) and thus is able to trace the behaviour of chemical species within a given geochemical environment. Since socio-biological behaviour of organisms may be disturbed at concentrations below the picogram per kg limit, the sensitivity and resolution of the applied analytical method must be very high.

In the case of trace organics various chromatographic separation methods and sensitive detectors (e.g. capillary gas chromatography with electron-capture detector or mass spectrometry, HPLC-fluorescence spectrophotometry) can solve these tasks. However, the direct determination of trace metal species is still in its infancy. Here, the widespread availability of atomic absorption spectrometry, which itself cannot solve this task, has possibly delayed the search for adequate chemical analyses and equilibrium calculation methods.

Since the already classic study of Tyler (1972) forwarded the hypothesis that heavy metals may reduce the productivity of terrestrial ecosystems, the distribution of trace metals has generally been given as total metal concentration within a compartment. However, in order to explain their transport mechanisms, their chemodynamics at interfaces, or their toxicity, we need some information on their speciation (Florence, 1986).

Thomas *et al.* (1984), who studied the accumulation of airborne pollutants (polycyclic aromatic hydrocarbons, chlorinated hydrocarbons, and heavy metals) in various plant species and humus in southwest Sweden, circumvented this problem by applying multivariate statistical analysis and knowledge of the physicochemical properties of the individual chemicals in order to explain the fate of the trace metals within the ecosystem. They concluded from the principal component analysis and the differences of the distribution patterns of the individual pollutants that the enrichment of the polycyclic aromatic hydrocarbons, Pb and V is caused by their deposition adsorbed onto small atmospheric particles. The reason for the variable enrichment is the different specific surface of individual plant species. The other metals (Zn, Cd, and Cu) are characterized by lower retention capacities of the vegetation and a selective uptake mechanism for the fungi species analysed. On the other hand the authors were not of the opinion that an accumulation of pollutants through the root system is important.

Chlorinated hydrocarbons can be enriched more effectively by those plant species that contain fatty and aromatic substances in which these pollutants are soluble.

The limitations of such studies are evident when we ask, for those metal species, their activities and their chemodynamics that would affect an individual population or the biocoenosis.

In the late 1970s and early 1980s it became evident that aluminium appeared to be toxic to fish (Muniz and Leivestad, 1979; Baker and Schofield, 1982), and possibly to plant roots (Ulrich *et al.*, 1980; Bauch, 1983). It soon became evident that the extent of toxicity was dependent on the form of aqueous aluminium. Several workers developed procedures for the fractionation of aqueous aluminum (e.g. Barnes, 1975; Driscoll, 1984). Since that time a wide range of studies related to the 'acid rain' phenomenon have presented detailed information (detailed, compared with other trace metals, e.g. Bourg and Védy, 1986) on the chemodynamics of aluminium species in terrestrial ecosystems (e.g. Litaor, 1987; Driscoll, 1985). Further, the effects of hydrologic pathways on the mobilization and speciation of aluminium from soils to receiving waters have been studied (e.g. Cozzarelli *et al.*, 1987).

By summarizing this information it appears that in porewater and groundwater (Fig. 6.2) mononuclear Al occurs as aquo Al, OH^-–, F^-–, SO_4^{2-}– and organic complexes. The inorganic Al species seem to increase with decrease of the pH of the solution.

The studies of aluminium species, however, focus mainly on the soil and aquifer compartments of terrestrial ecosystems and the data on transfer and chemodynamics of aluminium species within and between other compartments are far less conclusive.

Few attempts have been made to measure the distribution of organic micropollutants within a terrestrial ecosystem (Herrmann, 1987a). With the

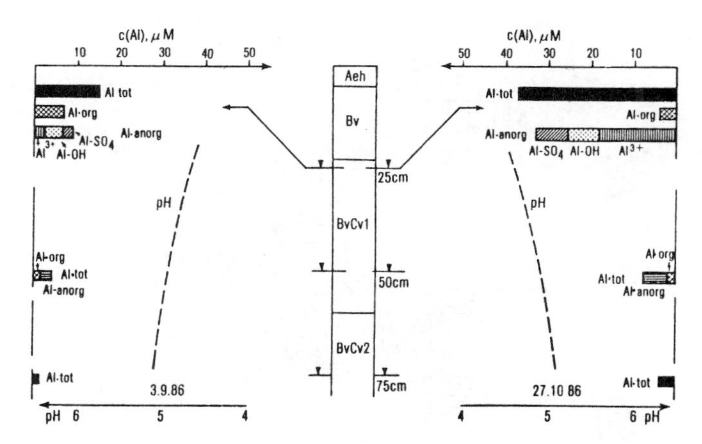

Figure 6.2 Variations with depth of concentrations of aluminium species and pH values in soil water of a podsolic brown earth/Fichtelgebirge, Upper Franconia. Maximum and minimum concentrations of a time-series from August to December 1986 (data from Stenzel and Herrmann, 1988).

help of the following example (Herrmann, 1987b) I focus on the potentials and limits of determining environmental chemodynamics of organic micropollutants.

At the end of the 1970s about 63 000 chemicals were in common use, some of which are hazardous bioavailable micropollutants that may leak into terrestrial ecosystems (Maugh, 1978). Thus, compared with inorganic micropollutants, such as trace metal species, we are now confronted with a vast number of organic micropollutants of which an unknown number might also act together in a synergistic way. I demonstrate this with two chromatograms of halogenated hydrocarbons from the headwater of the Püttlach stream, in Upper Franconia, West Germany (Fig. 6.3). Since we do not know the mass spectra of many organic micropollutants, and are not able to obtain calibration standards, it is possible to identify only a few of the peaks in high-resolution capillary gas chromatography. Therefore, we are also unable to perform any effect studies with these unknown substances.

However, as long as our aim is to study the transport behaviour of organic micropollutants within the ecosystem, one possible way of dealing with the great number of pollutants is to choose model substances with a wide range of physicochemical properties and to study their behaviour within the ecosystem. Then, the environmental behaviour of other substances can be estimated by means of the similarity of their physicochemical properties to those of a model substance. In order to follow this approach we need to implement a measuring system in the field and in the laboratory that is capable of coping with the analysis of a wide range of chemicals (Herrmann,

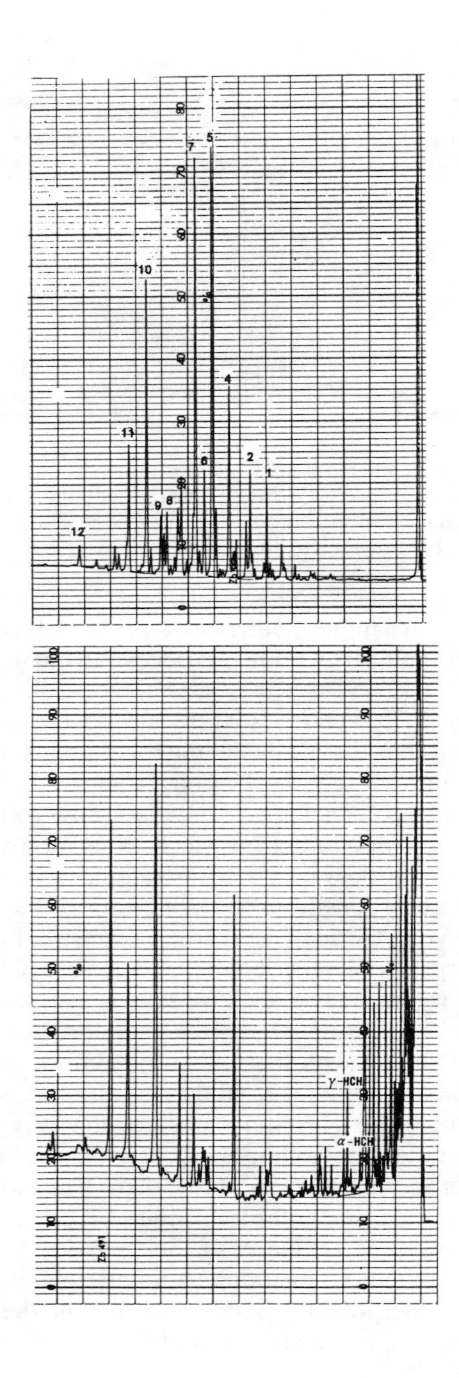

1
2
4
5
6
7
8
9
10
11
12
γ-HCH
α-HCH

1987b). Since the environmental transfer of organic micropollutants is also a function of the dynamics and structure of the ecosystem, an appropriate measuring system has also to take this into consideration (e.g. the conveying processes: wind, water, particles, or organisms). Although it is a formidable task to analyse these transfer processes in relation to the transport of organic micropollutants in an ecosystem, there are even more difficult problems for chemical analysis within these conveying processes without disturbing the process itself.

This might be illustrated with the example of the difficulty of measuring the flux of organic micropollutants in the soil under natural conditions. Extracting sufficient soil water for the analysis disturbs the water flow and changes sorption and intraphase transport processes between adsorbed, dissolved, and gaseous states of the trace organic in an unpredictable way. Thus, the measuring system disturbs the function of the ecosystem itself.

What can we learn from a study of environmental transfer of organic micropollutants within a hillslope ecosystem connected to a river? To illustrate how we might do this I summarize some results from Herrmann *et al.* (1986), and point out their limits in predicting the exposure concentrations that might realistically exist within individual compartments.

The concentrations and short-term fluxes of model substances of non-polar organic micropollutants were measured within the compartments atmospheric boundary layer, soil aquifer, river, and aquatic food web (Table 6.1). The mobility of these organic micropollutants within the ecosystem is a function of their physicochemical properties and the possibility of being transported in the gaseous phase, dissolved phase or adsorbed on particles (Figs 6.4 and 6.5). The results of our measurements and model computations indicate a decrease of mobility within the ecosystem from trichloroethylene over γ–HCH to 3,4-benzopyrene. The different physicochemical properties of the model substances bring about large amounts of volatile chlorinated hydrocarbons in the atmosphere boundary layer, whereas γ–HCH and 3,4-benzopyrene are predominantly adsorbed within the soil.

At this point I wish to stress that we cannot draw conclusions in respect to degradation processes, such as hydrolysis or photolysis, from this sort of study; I would like to stress further that transport processes and chemodynamics in microenvironments, e.g. in soil pores or within the boundary

---

Figure 6.3 (opposite) Capillary gas chromatograms with electron-capture detection of halogenated hydrocarbons extracted by a hexane/acetone mixture (1:1) from the same headwater sample of the Püttlach stream, Upper Franconia, with pre-chromatography by silica gel and aluminium oxide. Trace 1 shows the pattern after the first elution with hexane, trace 2 the pattern after the second elution with a mixture of diethyl ether and petroleum ether. Out of the first trace we identified 13 chemicals, mostly PCBs and HCB, out of the second trace, the HCH isomers.

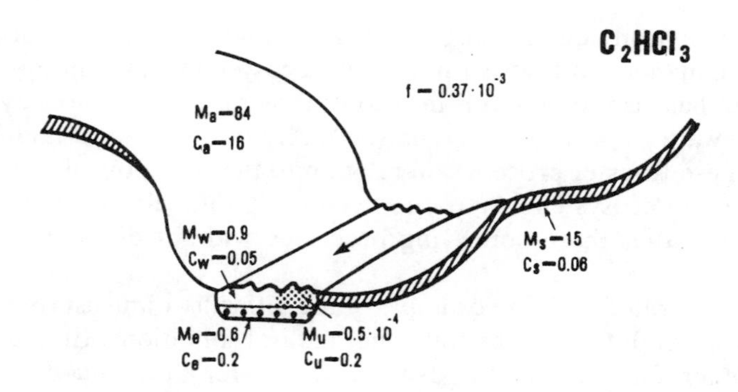

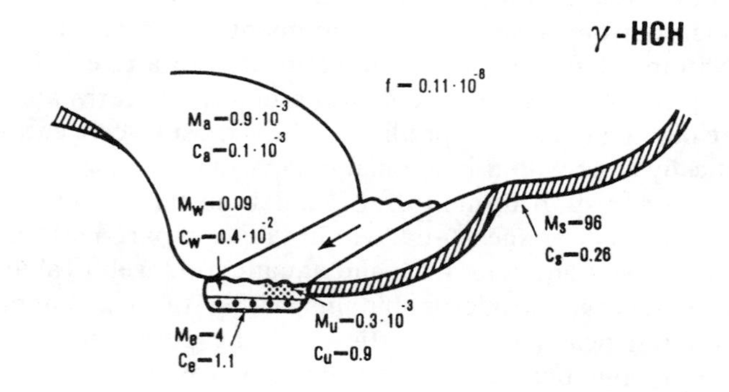

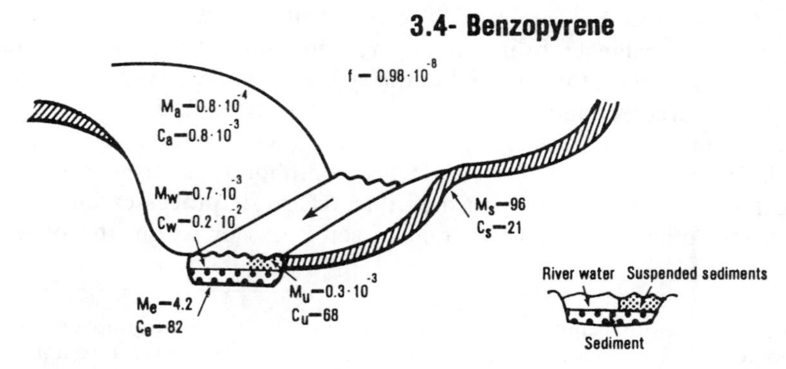

Figure 6.4 Equilibrium partitioning of organic micropollutants into various compartments of the hillslope and valley ecosystem. M: mass percentage (%); c: concentration (μg kg^{-1}); f: fugacity (Pa), can be viewed as the 'escaping tendency' that a substance exerts from any given phase (e.g. Mackay and Paterson, 1982); indices: a, air; w, water; s, soil; e, sediment; n, suspended sediment (from Herrmann, 1987a).

Table 6.1 Physicochemical properties of model substances (from Herrmann, 1987b)

Micropollutant	Tetrachloro-ethylene, C_2Cl_4	γ-Hexachloro-cyclohexane, $C_6H_6Cl_6$	3,4-Benzo-pyrene, $C_{20}H_{12}$
Structural formula	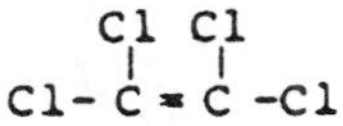 Cl Cl / Cl- C = C -Cl	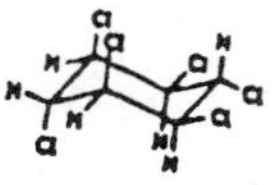	
Molecular mass ($g\ mol^{-1}$)	166	291	252
Vapour pressure, 298 K (Pa)	1.9×10^3	2.2×10^{-3}	1.9×10^{-6}
Henry's constant 298 K ($Pa\ m^3\ mol^{-1}$)	2.1×10^3	0.11	0.48
lg k_{ow}	2.6	4.1	6.2
Solubility in water, 298 K ($mg\ l^{-1}$)	150	6	0.99×10^{-3}

layer of the stomata—however important they may be—are out of the reach of available sampling techniques.

We can show by means of the non-stationary conceptional model of Mackay (Herrmann, 1987b), which is based on the assumption of equilibrium distribution of non-reacting organic micropollutants within and between the compartments (Fig. 6.6), that the time-series of selected model substances vary in different ways. As some trace organics fluctuate strongly, organisms or communities exposed to these trace pollutants must be able to adapt to time-variant stress.

Simulations like these mask quite a few difficulties: the change of Henry's law constant with changing temperature is, with few exceptions, not known. Thus the partitioning between air and other compartments is uncertain. Further, the exchange with the surrounding ecosystems and the atmosphere is rarely measured. The model predicts a fast flux, e.g. for polycyclic aromatic hydrocarbons from water to soil matrix or sediments, which we do not find by measurement. We attribute this discrepancy to the influence of dissolved organic carbon on the solubility of hydrophobic trace organics.

As the ecosystem is small with a considerable flow of air and water across the boundaries, and as there are uncertainties about the concentrations within these conveying media, validation is difficult. For example, water and suspended sediments are renewed every hour, and air every four minutes. As a result, concentrations in these compartments are dependent only on inflow concentrations, and rates of degradation and transfer are unimportant (and undeterminable). However, predictions of concentrations in biota, sediments, and soils show reasonable results (Trapp and Herrmann, 1986).

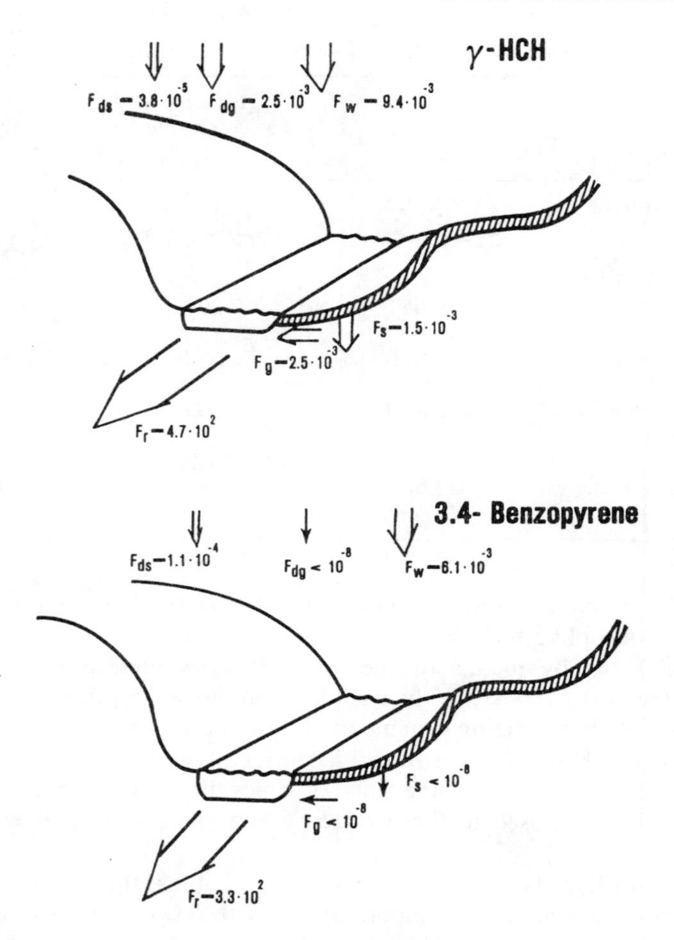

Figure 6.5 Flux diagrams for γ-HCH and 3,4-benzopyrene in the hillslope–valley ecosystem. Fluxes F in ng m^{-2} s^{-1}; indices: ds, dry adsorbed deposition; dg, dry gaseous deposition; w, wet deposition; s, soil water; g, groundwater; r, river water (from Herrmann, 1987a).

6.4 CONCLUSIONS

Given the temporal and spatial variation of the transfer processes, and the unstable nature of the processes involved, it seems impossible to continuously measure or model the environmental fate of organic and inorganic micropollutants with a high intercompartmental resolution in the near future. Even a sophisticated and comprehensive, uncontrolled ecosystem study would not allow us to assess the contribution of different chemodynamic processes or important transfer processes in microenvironments. Thus, I would like to

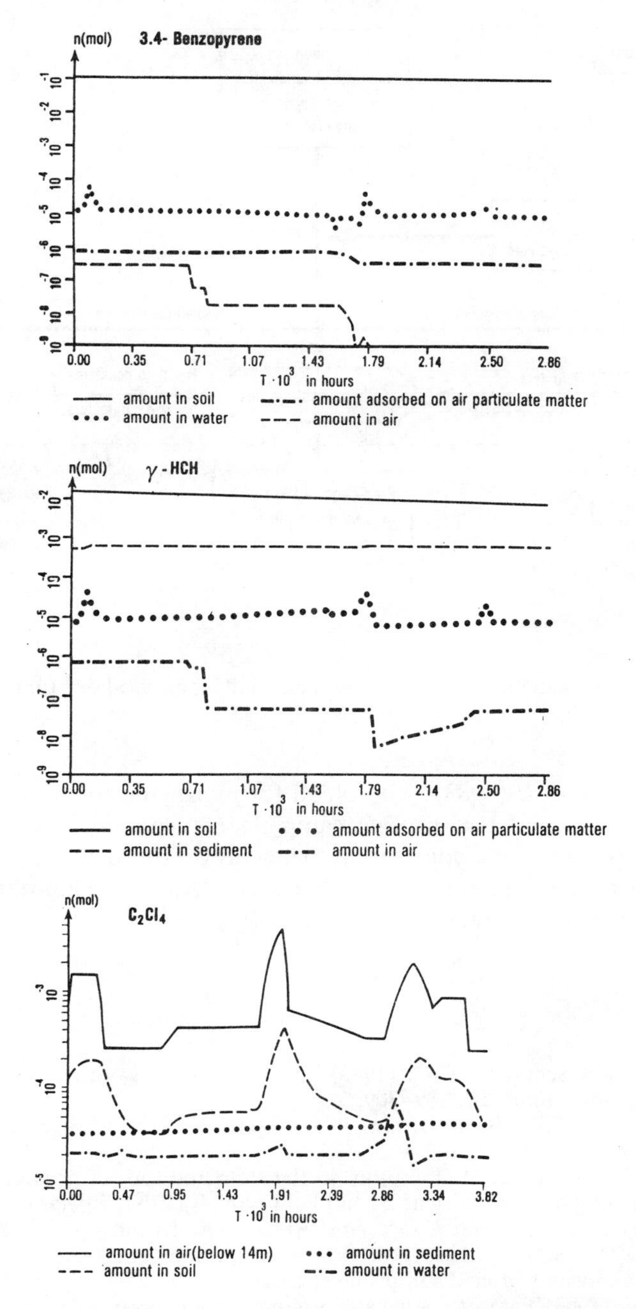

Figure 6.6 Time-series of compartmental distributions of model substances, simulated by a non-stationary Mackay model for the hillslope–valley ecosystem (redrawn after Herrmann, 1987b).

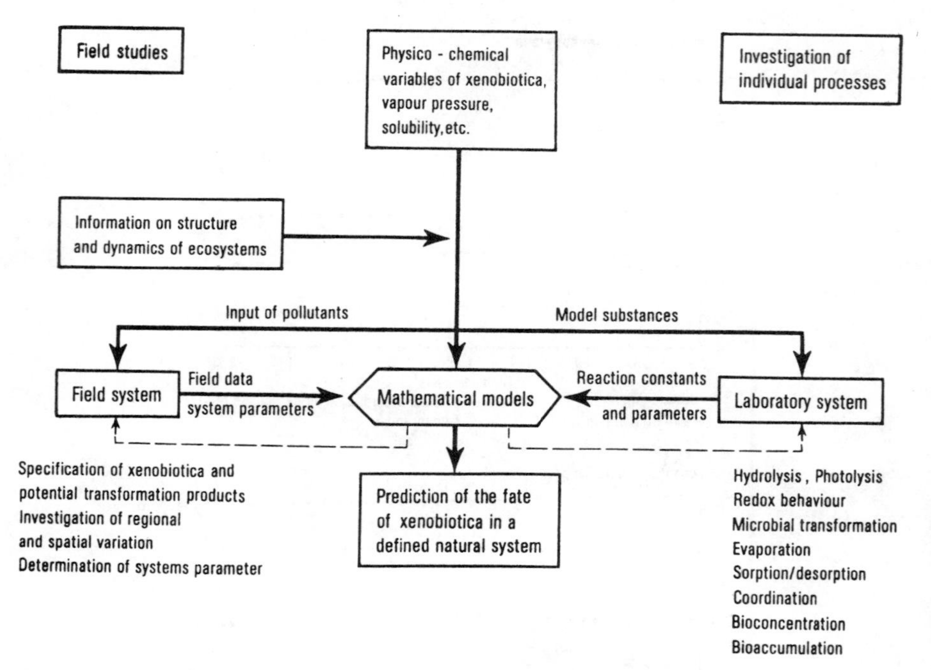

Figure 6.7 Assessment of environmental chemodynamics (after Stumm *et al.*, 1983).

follow Stumm *et al.* (1983), who plead for more concepts in environmental chemodynamics and less routine sampling in ecotoxicology and standard toxicity tests with single populations. They favour a procedure that combines field studies, investigation of individual processes in the laboratory and the use of mathematical models (Fig. 6.7).

REFERENCES

Baker, J.P. and Schofield, C.C. (1982) Aluminum toxicity to fish in acidic waters. *Water Air Soil Pollut.* **18**, 289–309.

Barnes, R.B. (1975) The determination of specific forms of aluminum in natural waters. *Chem. Geol.* **15**, 177–91.

Bauch, J. (1983) Biological alterations in the stem and root of fir and spruce due to pollution influence. In: Ulrich, B. and Pankrath, J. (Eds), *Effects of Accumulation of Air Pollutants in Forest Ecosystems*, Reidel, Dordrecht, pp. 377–86.

Bourg, A.C.M. and Védy, J.C. (1986) Expected speciation of dissolved trace metals in gravitational water of acid soil profiles. *Geoderma* **38**, 279–92.

Burns, L.A. (1983) Fate of chemicals in aquatic systems: process models and computer codes. In: Swann, R.L. and Eschenroeder, A. (Eds), *Fate of Chemicals in the*

Environment; Compartmental and Multimedia Models for Prediction, ACS Symposium Series 225, American Chemical Society, Washington, DC, pp. 25–40.

Cozzarelli, I.M., Herman, J.S. and Parnell, R.A., Jr (1987) The mobilization of aluminum in a natural soil system: effects of hydrologic pathways. *Water Resources Res.* **23**, 859–74.

Driscoll, C.T. (1984) A procedure for the fractionation of aqueous aluminum in dilute acid waters. *Int. J. Environ. Anal. Chem.* **16**, 267–84.

Driscoll, C.T. (1985). Aluminum in acid surface waters: chemistry, transport, and effects. *Environ. Health Perspect.* **63**, 93–104.

Florence, T.M. (1986) Electrochemical approaches to trace element speciation in waters—a review. *Analyst* **111**, 489–505.

Herrmann, R. (1987a) Environmental transfer of some organic micropollutants. In: Schulze, E.-D. and Zwölfer, H. (Eds), *Potentials and Limitations of Ecosystem Analysis*, Ecological Studies, **61**, Springer, Berlin, pp. 68–99.

Herrmann, R. (1987b) Chemodynamik und Transport von organischen Umweltchemikalien in verschiedenen Kompartimenten eines Talökosystems. *Gewässerschutz, Wasser, Abwasser* **100**, 229–301.

Herrmann, R., Eiden, R., Horn, R. and Zwölfer, H. (1986) Vergleichende Untersuchungen zur Mobilität von organischen Umweltchemikalien in und zwischen verschiedenen Kompartimenten eines Ökosystems. *Forschungsbericht* 10604023, Umweltbundesamt, Berlin.

Lassiter, R.R., Baughman, G.L. and Burns, L.A. (1979) Fate of toxic organic substances in the aquatic environment. In: Jorgensen, S.E. (Ed.), *State-of-the-art in Ecological Modelling*, Pergamon Press, Oxford and New York, pp. 219–46.

Litaor, M.I. (1987) Aluminum chemistry: fractionation, speciation, and mineral equilibria of soil interstitial waters of an alpine watershed, *Front Range, Colorado. Geochim. Cosmochim. Acta* **51**, 1285–95.

Mackay, D. and Paterson, S. (1982) Fugacity revisited—the fugacity approach to environmental transport. *Environ. Sci. Technol.* **16**, 654–60A.

Maugh, T.H. (1978) Chemicals: How many are there? *Science* **199**, 162.

Muniz, I.P. and Leivestad, H. (1979) Toxic effects of aluminum on the brown trout, *Salmo trutta* L. In: Drablos, D. and Tollan, A. (Eds), *Ecological Impact of Acid Precipitation*, SNSF Project, Oslo, pp. 268–9.

Stenzel, A. and Herrmann, R. (1988) Verhalten verschiedener Aluminiumspezies im Fluß- und Bodenwasser des Fichtelgebirges. *Deutsch, Gewässerkdl. Mitt.* **32**, 2–7.

Stumm, W., Schwarzenbach, R. and Sigg, L. (1983) Von der Umweltanalytik zur Ökotoxikologie—ein Plädoyer für mehr Konzepte und weniger Routinemessungen. *Angew. Chem.* **95**, 345–55.

Thomas, W., Rühling, Å. and Simon, H. (1984) Accumulation of airborne pollutants (PAH, chlorinated hydrocarbons, heavy metals) in various plant species and humus. *Environ. Pollut. Ser. A* **36**, 295–310.

Trapp, S. and Herrmann, R. (1986) Validation of fugacity models in the river Rotmain valley. Proceedings, 'Environmental modelling for priority setting among existing chemicals'. Workshop 11–13, November 1985, GSF, München-Neuherberg, pp. 253–67.

Tyler, G. (1972) Heavy metals pollute nature, may reduce productivity. *Ambio* **1**, 52–9.

Ulrich, B., Mayer, R. and Khanna, P.K. (1980). Chemical changes due to acid precipitation in a loess-derived soil in Central Europe. *Soil Sci.* **130**, 193–9.

7 Whole-lake Experiments at the Experimental Lakes Area

D.W. SCHINDLER
Department of Zoology, University of Alberta, Edmonton, Alberta, Canada T6G 2N6

7.1 INTRODUCTION

Man's effects on natural ecosystems are now so numerous and far-reaching that it is often difficult to deduce which of the many insults experienced at one time by a particular ecosystem are causing it to degrade. For example, legislation controlling eutrophication was delayed for several years by uncertainty about which of the many nutrients added to lakes by humans was causing the problem. Similarly, controversy concerns the factors that cause lakes to acidify, and the observed decline in European forests in recent years has been variously attributed to acid rain, ozone, exhaustion of soil micronutrients, overfertilization with nitrogen, natural climate changes, or a combination of the above. The potential interactions of known and perhaps yet unanticipated causes render it impossible to deduce from the results of traditional monitoring approaches which of man's activities need alteration. The above-noted forest problem is clearly one that could have benefitted from ecosystem-scale experiments in the era before multiple air pollution stresses were widespread, as Wright illustrates (this volume) (see also Wright *et al.*, 1988). Unfortunately, man's modification of the biosphere began long before the importance of ecosystem-scale experimentation was recognized, and suitable sites for valuable experiments can no longer be obtained in many areas.

Experimental modification of whole ecosystems is not new. It has been done for millennia, if agricultural and fishpond work is included. However, incorporating detailed studies of ecosystem-scale processes and changes in biological communities as a part of such large-scale manipulations is relatively new. Indeed, even today very few large-scale experiments include detailed, long-term analyses of biotic communities, community energetics, and biogeochemical cycles. For this reason I shall confine my analysis to chemical experiments at the Experimental Lakes Area (ELA), where we have attempted to incorporate key responses of both communities and ecosystems in our study designs. Other chapters in this volume will provide information on

Ecosystem Experiments. Edited by H.A. Mooney *et al.*

other types of deliberate ecosystem manipulations, including the direct manipulation of food chains, at which we at ELA are relative novices.

7.2 A BRIEF SUMMARY OF ELA EXPERIMENTS

In the past 20 years we have devised experiments to elucidate the effects of a number of human-caused problems in lakes. Whole lake experiments have included eutrophication; acidification; pollution with radioactive materials and trace metals; and, very recently, a beginning in food chain manipulations. We have also examined the effects of watershed changes on lakes, either by deliberately modifying watersheds (for example, the wetland acidifications of Bayley *et al.*, 1987 and Bayley and Schindler, this volume) and the clearcutting of forests (Nicolson, 1975). We have also studied natural 'experiments' in the case of windstorms and forest fires (Schindler *et al.*, 1980; Bayley and Schindler, 1987 and this volume). A total list of our deliberate experiments is given in Table 7.1. I have reviewed the results of these elsewhere (Schindler, 1988a). Here I describe features that rendered our ecosystem experiments uniquely valuable, and review why certain approaches to ecosystem experiments, as well as smaller scales of experimentation that we have employed, were successful at predicting ecosystem responses while others were not.

7.2.1 DESIGN OF EXPERIMENTS

One noteworthy feature of ELA ecosystem-scale experiments is that we have seldom tried to mimic a pollution problem in every detail. Instead, we have simplified our experimental treatments, reducing the number of possible alternatives that could explain any observed disruptions in the community or ecosystem. Otherwise, interpretation of our results would pose many of the same problems encountered in traditional monitoring approaches to deducing environmental problems. For example, by deliberately omitting carbon and trace elements from fertilizers in many of our eutrophication experiments, we were able to show that these elements played a much less important role than phosphorus and nitrogen in eutrophication (Schindler, 1974; Schindler *et al.*, 1972, 1973). By varying the ratio of nitrogen to phosphorus in our addition experiments, instead of confining the ratios to those found in sewage, we were able to elucidate the roles of atmospheric gases and blue-green dominance in the phytoplankton in balancing the nitrogen cycle of lakes (Schindler, 1977). Likewise, by adding acid without aluminum in our whole-lake experiments, we showed that most of the early biological effects of acidification are caused by hydrogen ion alone (Schindler *et al.*, 1985). Later experiments confirmed the secondary role of aluminum in acidification damage to lacustrine organisms (Playle, 1985). In an ongoing experiment we are adding sulfuric and

Table 7.1 Whole lake ecosystem experiments involving additions of chemicals at the Experimental Lakes Area, 1969–90

Site	Date	Treatment	Mesocosm experiments
Eutrophication experiments			
Lake 227	1969–present	P,N	×
Lake 226NE	1973–80	P,N,C	×
Lake 226SW	1973–80	N,C	
Lake 304	1971–72	P,N,C	×
	1973–74	N,C	
	1975–76	P,N	
Lake 303	1975–76	P,N	×
Lake 261	1973–76	P	
Lake 230	1974–75	P,N (Winter)	
Lake 302N	1972–76, 1978	P,N,C (Hypolimnion)	
Acidification experiments			
Lake 223	1976–present	H_2SO_4	×
Lake 302S	1982–present	H_2SO_4	×
Lake 302N	1982–86	HNO_3	
Lake 114	1979–86	H_2SO_4	×
	1984	$(Al)_2SO_4$	
239 Wetland	1983–present	$H_2SO_4 + HNO_3$	
Radioactive materials			
Lake 227	1970	^{226}Ra	×
	1975	^{3}H	×
	1978	^{32}P	×
Lake 261	1971	^{226}Ra	
Lake 224	1976	^{226}Ra, ^{14}C, ^{3}H, ^{75}Se, ^{203}Hg, ^{134}Cs ^{59}Fe, ^{65}Zn, ^{60}Co	×
Lake 226NE and SW	1977	^{75}Se, ^{203}Hg, ^{134}Cs, ^{59}Fe, ^{59}Fe, ^{65}Zn, ^{60}Co, ^{89}Sr	×
	1978	^{14}C	×
239 Wetlands	1984	^{3}H	
Trace metals			
Lake 382	1988–present	Cd	×
Chlorinated organic compounds			
Lake 260	1989–present	Furans, toxa phene (injected into fish)	

nitric acids separately to lakes, in an attempt to resolve the debate over the relative merits of controlling emissions of sulfur and nitrogen oxides (Rudd *et al.*, 1990; Kelly *et al.*, 1990).

7.2.2 THE ELEMENT OF SURPRISE

Many of our results at ELA were not predictable before the experiments were begun. For example, no one had predicted the key role that atmospheric exchange of carbon and nitrogen would play in the eutrophication of lakes (Schindler *et al.*, 1972, Schindler, 1977). Before our experiment in Lake 227, ignoring atmospheric sources of nutrients had caused many to conclude that lakes must be limited by carbon (reviewed by Schindler, 1971). Later, we elucidated a similar role for atmospheric nitrogen, in allowing nitrogen-deficient lakes to eutrophy when excess phosphorus was supplied. In the case of nitrogen, a switch to nitrogen-fixing blue-greens was also necessary to facilitate the fixation of atmospheric nitrogen (Schindler, 1974, 1977; Schindler *et al.*, 1985). The unimportance of phosphorus return from sediments of fertilized Precambrian Shield lakes was also unexpected (Schindler *et al.*, 1977, 1987; Levine *et al.*, 1986).

In our acidification studies there was no basis for predicting that acidification would affect fisheries via food chains as well as by direct toxicity to fishes (Schindler *et al.*, 1985); indeed, except for our studies at ELA, the question of food chain effects has been ignored in acid rain research. The complicated sequences of factors that damaged some populations of organisms in our experiments would have been difficult to foresee, or to examine using smaller-scale experiments. For example, the decline and extirpation of crayfish populations appeared to be caused by a combination of physiological factors, parasitism, and reproductive failures (Schindler *et al.*, 1985; France, 1987; Davies, 1989).

In the 1970s it was widely believed that the buffering of lakes was controlled by geological processes, chiefly in the terrestrial watersheds of lakes. These sources of buffering were believed to be exhausted by acid rain, causing lakes to become irrecoverably acidic. Our experiments showed that not only were several *in situ* sources of buffering important, but that some of these were biologically mediated (bacterial reduction of sulfate and nitrate) instead of being strictly geochemical processes. These sources increased, rather than decreased, in response to increased acidification (Schindler, 1980, 1986; Cook and Schindler, 1983; Cook *et al.*, 1986). Also contrary to popular belief, the most important sites of sulfate reduction and denitrification (and therefore sources of alkalinity) were in epilimnion sediments, rather than in the hypolimnions of lakes (Cook *et al.*, 1986; Kelly and Rudd, 1984). Smaller-scale methods were calibrated in whole-lake experiments, then applied elsewhere, revealing that the alkalinity-generating processes occurred in all acidifying lakes (Rudd *et al.*, 1986). Such testing and calibration of smaller-scale methods is an important function for whole-ecosystem experiments, as I shall discuss later. As the result of both whole-ecosystem experiments and investigations elsewhere using the calibrated smaller-scale methods, it was possible to develop a single model to predict the roles of both nitrate and sulfate removal in retarding acidification (Kelly *et al.*, 1987). The element of unpredictability often confounds our present ability to do environmental forecasting at several scales (Holling, 1985). Ecosystem experiments can often expose unforeseen factors, substantially reducing the number of unforeseen surprises.

7.2.3 DURATION OF EXPERIMENTS

The importance of maintaining experiments for several years has become very apparent, for it allows us to estimate how long the modified ecosystem requires to approach a new steady state, or at least some approximation thereof. Our longest-running experiment, the eutrophication of Lake 227, is in its 22nd year in 1990. As a result we have been able to address such long-term questions as saturation of surface sediments with nutrients, and the approach of nutrient biogeochemical cycles to a new steady state. In this case we have found that the nitrogen cycle of the lake is still changing, even though nitrogen loading has not changed since 1975 (Schindler *et al.*, 1987). In the acidification of Lake 223, now in its 13th year, we have seen lags of several years in the response of long-lived species to chemical change. Recruitment failure is the most common of all early biological problems, and long-lived populations may survive for years or decades after reproduction ceases. Time lags of several years were also found to occur in the responses of populations of fishes to eutrophication and acidification (Mills and Chalanchuk, 1987; Mills *et al.*, 1987 and unpublished).

Even among microbial populations, response times can be substantial. Rudd *et al.* (1988) showed that aquatic nitrifying bacteria did not recover fully from acidification for one year after suitable pH values returned. Thus, neither biogeochemical cycles nor biotic communities necessarily adjust quickly to new stresses, and critical experiments must usually be of several years' duration.

7.3 LARGER EXPERIMENTAL SCALES—CAPITALIZATION ON DELIBERATE CHEMICAL PERTURBATIONS

Due to the other demands of human society, as well as high costs, it is impossible in most cases to deliberately perform experiments at scales larger than small ecosystems for scientific purposes alone. Yet, as mentioned above, there are likely to be many mesoscale features that change responses to perturbation somewhat. For example, larger lakes usually have more complex biotic communities that should confer greater resistance to perturbation via greater redundancy. Similarly, advective, as opposed to convective, mixing processes become increasingly important. Response times are often slower in larger ecosystems, especially if water renewal times are longer. As a result it may be necessary to devote special study to scaling factors before we can apply some of the results of experiments in small ecosystems to large ones. Large perturbations can actually change climate, as well as involving interactions between several ecosystem types. Usually, we must capitalize on environmental modifications done for other reasons—for example, reasons of economics or health—in order to address these scales. There are a few larger-scale 'experiments' of this sort that are quite informative. For example, economic slowdowns and emissions controls have caused the sulfur oxide emissions from Ontario sources to decrease substantially in the past decade. As predicted by ecologists, this has caused considerable chemical recovery in many lakes of the area (Dillon *et al.*, 1986 and unpublished data; Schindler, 1987). More interestingly, and less predictably, in Ontario lakes where detailed biological studies had been done during the high-deposition phase of the 'experiment' it has been possible to substantiate that a considerable biological recovery has occurred (Hutchinson and Havas, 1986; Dillon *et al.*, 1986; Keller and Pitblado, 1986; Gunn and Keller, 1990). Populations of fishes that had not reproduced for years resumed recruitment, showing that biological steady state with acid deposition was never reached. Information of this sort is a powerful incentive for politicians to act swiftly to control sources of air pollution, for there are still populations of fishes that might be saved from extirpation. Such considerations of biological steady state have not been addressed in the debate over the control of acid rain, although I have repeatedly mentioned the possibility that they might exist (Schindler, 1987, 1988a,b). Of course, one of the motives for this volume is the concern that man is conducting an uncontrolled experiment with the entire earth, without any concern for the necessary adjustments in scale that are necessary to apply what we know about smaller-scale ecology and geochemistry!

7.4 TYPES OF ECOSYSTEM RESPONSE TO EXPERIMENTAL PERTURBATION

In general, the earliest responses of ELA ecosystems to our experimentally imposed stresses have been the elimination of sensitive species (Schindler, 1987, 1990). Ecosystem functions such as primary production, respiration, and nutrient cycling are usually less vulnerable, apparently due to high

Table 7.2 A comparison of the sensitivity of several indicators of ecosystem stress to acidification of whole lakes with H_2SO_4. Data from studies in Lakes 223 amd 302S; from Schindler (1990)

pH	6.5	6.0	5.5	5.0
Sensitive species decline or disappear		—♦—	*Mysis*, *Pimephales*, *Hyalella*	
Fish cease reproducing		♦Fathead minnow ♦	Lake trout Pearl dace ♦	Sucker
Species diversity declines			♦Phytoplankton, chironomids	
Major food web disruptions		———♦———		
Fish mortalities increase		♦Pimephales	♦Lake trout	
Conspicuous algal mats			♦ Mougeotia, Zygnema	
Increase in average phytoplankton size			♦Dinoflagellates	
Phytoplankton production		No decline		
Winter respiratory declines			♦	
Ecosystem P/B		No change		
Ecosystem R/B increases			♦	
Ecosystem P/R decreases			♦	
Periphyton production decreases	♦			
Periphyton respiration increases	♦			
Periphyton P/R decreases	♦			
Σ Nitrogen increases			♦	
Nitrification ceases			♦	

redundancy and short generation times in the biotic communities that carry out key ecosystem functions. Indeed, I believe that it is in those ecological niches where redundancy is low, life cycles are short and the resident organisms are particularly sensitive to a given perturbation that must have first priority for ecosystem protection. It follows that good taxonomy and community ecology must be the underpinnings of any program to protect ecosystems, with studies of ecosystem function playing a secondary role. Once ecosystem functions have been altered the problem is very serious, as we can seen from the present perturbations of the global carbon cycle and ozone layer.

There are exceptions to the rule that aquatic populations and communities are more sensitive to perturbation than ecosystem functions. Some components of the chemical cycles of lakes are very sensitive. For example, alteration of the balance between periphyton production and respiration was among the most sensitive responses to acidification that we observed at ELA (Schindler, 1990; M.A. Turner, unpublished data; Table 7.2). Similarly, nitrification of ammonium was inhibited quite early in the acidification process (Rudd *et al.*, 1988). While in the context of whole ecosystem metabolism these are small in scale, the conclusion that species changes are more sensitive than changes in function must be a cautious and tentative one. If relatively minor changes in a biogeochemical cycle occur on a broad regional scale, serious consequences can result. Indeed, a scenario of this sort may be responsible for the increase in methane in the atmosphere.

7.5 CALIBRATION OF OTHER METHODS WITH ECOSYSTEM-SCALE EXPERIMENTS

Deliberate manipulation of whole ecosystems for scientific purposes is possible only in a few locations. As a result it is important to calibrate biomonitoring, toxicological, paleoecological, bioassay, mesocosm and other widely usable smaller-scale techniques, to ensure that they are sensitive enough and reliable enough to detect ecosystem dysfunction. I could furnish several examples of uncalibrated techniques that proved to give erroneous interpretations of ecosystem-scale events. Below, I discuss a few, based on our own experience.

Mesocosms (known to limnologists as limnocorrals, a term introduced by Wally Broecker upon viewing our large enclosures in Lake 227, in 1970) have yielded reliable results with various chemical and plankton experiments at ELA. Perhaps the most successful application of limnocorrals at ELA was to losses of trace metals and radionuclides. One simple model was developed that accurately predicted loss rates of a wide variety of elements, in mesocosms and lakes of several sizes (Santschi *et al.*, 1986). Experiments assessing the importance of various chemical and microbial contributors to alkalinity were also successful (Schiff and Anderson, 1987). Mesocosm studies of effects of acidification on plankton, periphyton and metals gave results that were similar to those observed in whole-lake experiments (Schindler, 1980; Müller, 1980; ELA staff, unpublished data).

Mesocosm studies of air–water and physical mixing processes were less successful. The wall of a mesocosm protects the surface of a lake from the wind to some degree, so that gas exchange is less than it is in an entire lake. The wave energy broken by the tube wall is instead imparted to the structure itself, so that bobbing and rocking of the limnocorral causes increased rates of vertical mixing (Hesslein and Quay, 1973; Quay, 1977).

Effects of nutrients on the onset of blue-green algal blooms proved difficult to simulate in mesocosms. It was necessary to redesign experiments intended to simulate the nutrient conditions in lakes three times before important features of the whole lake were properly mimicked. For example, different relative rates of recycling of nitrogen and phosphorus from the surface and subsurface of epilimnion sediments, mesocosm walls, and through thermoclines greatly altered the species composition of phytoplankton in individual corrals. Proper scaling of sedimentation and return from sediments was the most difficult factor to simulate, although problems were eventually overcome (Levine, 1983).

Temporal scales also impose limitations on mesocosm applications. Periphyton growth on the walls of limnocorrals can allow this community to dominate results to an unrealistic degree after a few weeks (Levine, 1983). Exclusion of incoming nutrients and larger organisms can also cause results to differ from those of whole lakes.

Experimental manipulations of higher trophic levels were very unsuccessful in mesocosms at ELA. The mesocosms were regarded as convenient 'lunch stands' by otters and eagles, the top predators in ELA lakes. Once again, scale was an important consideration. Even a 10 m diameter mesocosm is far too small to accommodate lake trout, which commonly have densities less than 1/1000 m^2, and which visit a variety of habitat types in the course of their seasonal movements and life cycles. Crayfish were impossible to confine in mesocosms open to natural sediments. They simply burrowed under the walls and escaped (I. Davies, unpublished data), or were eaten by otters that climbed into the enclosures.

Even in mesocosm experiments to investigate the dynamics of zooplankton and small zooplanktivores, accidental inclusion or omissions of predators, and proper scaling of predator and prey densities, have posed a few problems (M. Vanni, unpublished data). In summary, there are practical upper limits to the scales, durations, and food chain complexities that can be successfully included in mesocosm experiments, and the limits must be known in order to utilize mesocosms properly. After numerous disappointing comparisons of mesocosms and manipulated whole lakes, I would be reluctant to use uncalibrated mesocosm experiments as a direct guide to manage lakes. As alluded to earlier, the same considerations undoubtedly apply when we extrapolate from small ecosystems to larger scales. Such investigations are under way. For example, E. Fee and R.E. Hecky are comparing the temporal variability of phytoplankton production and other variables in lakes of northwestern Ontario spanning five orders of magnitude in size, but having similar rates of water renewal. After three years of study it appears that there is little difference in the inherent variability of the systems with respect to phytoplankton production, i.e. the small lakes are excellent models for the larger ones (E.J. Fee, personal communication).

Ecosystems can be utilized to design and test effective monitoring methods. They can illustrate variables to monitor that are sensitive indicators of stress, or indicators of particular types of stress. Frequently, general management strategies can be more thoroughly tested in controlled ecosystem experiments than by either smaller-scale experiments or large-scale monitoring approaches.

At still smaller scales, bottle experiments have even greater problems than mesocosms, as might be expected simply from considerations of scale and complexity. So-called nutrient bioassays have been widely used in an effort to deduce which nutrients would be the ones to control to halt or reverse eutrophication. We have found that nutrient limitations in bottles often reflect the *results* of perturbation, rather than the fundamental *causes* (Schindler, 1988a; see also Hecky and Kilham, 1988). Laboratory toxicity studies of the LC-50 type also appear to underestimate the sensitivity of organisms, usually because they do not necessarily include the most sensitive stages of the life cycle. Often they are also run using standard test organisms, rather than the species of actual concern.

Interpretations from paleoecological studies also require calibration. Warwick (1980) studied the chironomid assemblages that accompanied man's settlement of the Bay of Quinte, on the northern edge of Lake Ontario. He expected to find changes similar to those that had accompanied eutrophication in Europe (Brundin, 1956). Instead, the assemblage changed to resemble that of a more oligotrophic lake. This was puzzling, for an increasing human population should have meant higher inputs of nutrients. More detailed studies of the sediments and the history of the basin revealed that low-nutrient mineral soils eroded from deforested watersheds had diluted the nutrient-rich autochthonous sediments, 'fooling' the microfossils into 'believing' that the lake had become less productive! In another example, Davidson (1984) studied the diatom fossils in Lake 223 at ELA as we experimentally acidified the lake. Paleoecological interpretations indicated that the original pH of the lake was 6.0, not 6.5–6.8 as we actually measured it prior to beginning the experiment. They also predicted that acidification had begun in the mid-1960s, not in 1976 when we actually began to add acid. This latter error appears to be caused by the 'bioturbation' of near-surface sediments. Paleoecological methods are usually calibrated by correlating biological assemblages of fossil-producing organisms in nearby lakes with the variable that is believed to have caused the fossils in a particular lake to change over time. For example, diatom frustrules are often used to deduce changes in pH, and chironomid assemblages have been used to predict eutrophication. Such methods are based on the assumption that the contemporary assemblages studied are all responding primarily to a single driving variable in the same fashion that they did in times past, and that the driving variable is in all cases the one of paleoecological interest. My limited experience in paleoecology, primarily through supervising the above two studies, has led me to believe that paleoecological assumptions should also be subjected to calibration in experimental ecosystems.

7.6 LAKES AS ARCHIVES

Lakes also have a number of features that make them valuable for detecting past changes in terrestrial and wetlands in their watersheds. They act as enormous 'filters' for detecting and storing information about environmental insults in surrounding ecosystems on a number of temporal and spatial scales. For example, lake sediments offer a convenient 'library' recording the temporal histories of pollutants such as trace metals and organic pollutants, of both local and very distant origin. They also record the history of perturbations to the watershed, such as land-use changes or fires. This they do by preserving in sediments the remains of fossil-producing organisms, both those from the lake itself and from the watershed and nearby ecosystems. Pollen, seeds, and other structures resistant to decomposition are among the useful fossils found in dateable sediments. Sediments can be aged by using the decay of radioisotopes such as C-14 or Pb-210, or in exceptional cases by counting varves (visible bands indicating annual increments of sedimentation). Paleoecological studies dovetail nicely with long-term monitoring to give a long record of the natural behavior of ecosystems, helping us to deduce whether observed phenomena are natural or are responses to human activities. It is therefore important that the sites of whole-ecosystem experiments be chosen for their paleoecological, as well as contemporary ecological, potential. For example, many of the lakes at ELA have hypolimnia that are anoxic for all or most of the year, minimizing the mixing of sediments by biota. As a result, fossils are preserved in almost the exact sequence that they fall to sediment surfaces, often allowing strata to be dated with a resolution of a few months. Of course, interpretations of changes in the catchments of lakes should be subjected to calibration procedures similar to those described above for lacustrine communities.

7.7 THE STUDY OF NATURAL VARIABILITY AS A COMPANION TO WHOLE-ECOSYSTEM EXPERIMENTATION

In the early days of experimentation at ELA we did not concern ourselves much with detailed long-term studies. Indeed, the planned life of the project at its present site was only five years. We performed experiments that were sledgehammer blows, i.e. our perturbations were so large that no-one with common sense could question the results, despite the fact that experiments were unreplicated.

However, we had to change this strategy in our later studies of acid rain and other air pollutants. The effects of these on lakes had not been as well characterized by past research as the effects of nutrients, and were likely to be chronic, but subtle. Fortunately, by the time we performed these experiments, we had long records characterizing at least some of the features of natural, unperturbed ecosystems, and we were able to develop methods to distinguish between natural variation and subtle changes brought about by our experimental stresses (Kasian and Campbell, in review; Schindler *et al*, unpublished). Even though the ELA area contains thousands of available lakes, it is impossible to choose lakes that are similar enough to design an ANOVA-style experiment of the sort that can be done in bottles, mesocosms, or agricultural plots. The best we can do is to set confidence limits on natural variation in several unperturbed lakes, then calculate the probability that any larger variations in experimental systems might be due to chance. Also, a comparison of the similarity of responses among pseudoreplicated studies may give increased confidence in the veracity of their results. Concerns about pseudoreplication are certainly valid, but they should not preclude decisions or drawing conclusions in situations where pseudoreplication cannot be avoided. My experience has been that the judgement of an experienced and well-trained ecologist is often more trustworthy than statistical analysis, unless the experiment is very well designed indeed.

Long-term records, in both natural and experimentally modified basins, allow us to deduce how a number of long-term changes in weather or climate affect lakes and interact with experimental perturbations. For example, there has been a recent trend toward warmer, drier summers at ELA, particularly in the spring of the year. Mean water temperatures have increased by about 3°C in 20 years. Combined with a slight decrease in average precipitation, this has caused a dramatic decrease in the rates of streamflow and lake flushing. As a result, chemical concentrations and phytoplankton standing crops have increased (Schindler *et al.*, 1990). While it is too early to tell whether or not these changes are natural, or the result of greenhouse warming, they give us a preview of what changes greenhouse warming might cause in boreal lakes.

Some natural events also cause responses in lakes that are similar to human perturbations. Experimentally, we have found that both acidification and dry weather cause lakes to become clearer. In the former case, dissolved organic substances probably co-precipitate with aluminum (Effler *et al.*, 1985). Under drought conditions, lake water is replenished less rapidly, and photodegradation, flocculation and decomposition of dissolved organic matter in lakes cause a gradual clearing (Schindler *et al.*, 1991).

7.8 CAN ECOSYSTEM-SCALE EXPERIMENTS HELP IN PREDICTING THE EFFECTS OF GLOBAL CHANGE?

There are several types of experiments that might help to predict the effects of climate change and other large-scale perturbations on aquatic ecosystems. For example, warmer temperatures and reduced soil moisture in boreal watersheds would cause increased incidence and size of forest fires. The denudation of terrestrial watersheds would increase wind energy at the lake surface, causing deeper thermoclines, reducing habitat for cold stenotherms such as lake trout and *Mysis*, while increasing habitat for warm-water organisms such as pike, bass, and crayfish. Higher productivity might also intensify decomposition in the hypolimnion, causing further reductions in cold, oxic habitats. It is possible to deepen the thermocline of a lake simply by clearcutting its basin, and to test the effects of drought by adding nutrients, so that we might test these hypotheses now. Similarly, some of the hypothesized effects of permafrost melting on lake transparency and productivity could be simulated by removing the insulating surface organic layer from soils in the drainage basin of an ecosystem underlain by permafrost. The effects of reduced water renewal and lake level on a variety of physical, chemical, and biological phenomena would be relatively simple to study in whole-ecosystem experiments. To be most useful, the sites of experiments to predict the effects of climate change should be in areas where maximal effects are expected, for predictions could be verified sooner at such sites, so that results could be used to direct ameliorative efforts in areas changing less rapidly.

One other type of aquatic experiment has yielded results that should be useful in predicting the results of climate change. Thermal power plants based on small lakes often use the water as a coolant, heating lake temperatures by several degrees. In at least some respects this warming should cause changes similar to those induced by climatic warming. Patalas (1970) compared several characteristics of a reference and a warmed lake in Poland. He found that a temperature increase from 21–22°C to 26–27°C caused a twofold increase in primary production. While there were neither species changes nor changes in biomass of zooplankton after seven years, the turnover time of this group in the warm lake was only half as long as in the reference. Dominance also switched from copepods to cladocerans. Eloranta (1983) found that thermal warming caused an increase in the ice-free season of a Finnish lake. Higher oxygen deficits in the hypolimnion, decreased transparency (due to higher phytoplankton biomass), and higher zooplankton biomass were also observed.

In summary, it seems that incorporating more detailed studies of the effects of temperature increases into the plans for constructing future nuclear and thermal power plants would be fruitful.

7.9 EXPERIMENTAL RECOVERY OF ECOSYSTEMS

The most important part of an experiment relating to management problems is often the recovery of the system after it has been damaged. Yet it is often impossible to obtain funding for recovery studies. There is a strong tendency among managers to consider an ecological problem solved once the stress that has damaged an ecosystem has been removed, and to expect scientists to turn to other activities. This policy ignores such matters as the abilities of key organisms to reinvade damaged habitats once suitable conditions return, as well as the re-establishment of ecological balances and biogeochemical cycles. In ecological jargon the question is often posed as 'will the ecosystem return to its original steady state, will it return to a new, but stable state, or will it remain unstable (unrepaired) for a long time?' We have been able in some cases to study the recovery of a lake after our experimental stress has been removed. For example, the response of the phytoplankton, primary and secondary production, and nutrient chemistry of Lake 226 to termination of fertilization has been well documented (Shearer *et al.*, 1987; Findlay and Kasian, 1987; Mills *et al.*, 1987). Four years of studying the response of the Lake 223 ecosystem to increasing pH following eight years of acidification produced a number of surprises. For example, all species of phytoplankton recovered quickly. On the other hand, few if any of the species of Chironomidae eliminated by acidification have returned, even though they are winged insects which occupy other lakes within 0.5 km of Lake 223. Among fishes, white sucker and pearl dace resumed spawning in the first year that the pH was raised, although the suckers overreproduced so that individuals became very thin, had higher than normal mortality and poor recruitment during the first few years of recovery due to starvation induced by overpopulation. In brief, once an ecosystem has been destabilized, removing the perturbation does not necessarily mean that stresses on all natural members of the community will be immediately reduced.

7.10 A COMPARATIVE ANATOMY OF ECOSYSTEM EXPERIMENTS

Performing whole-ecosystem experiments will always be difficult and costly. For many ecosystem types it is difficult to obtain ecosystems that are free from conflicting perturbations or other human interests for periods long enough to perform the necessary work. Ecosystem-scale studies also require the participation of a wide variety of scientific experts. For these reasons it is necessary to design studies of a sort that will allow extrapolation to different ecosystems, countries, perturbations, and species of organisms. This is a long-term goal of ELA, but it also needs consideration attention elsewhere.

It is important to conduct analogous experiments in wetlands and on upland sites, for these may react differently from lakes, yet any changes will 'cascade' into lakes and streams in the watershed. For example, forest fires increase the concentrations of all chemicals in lakes or streams, and may significantly affect the buffering capacity (Bayley and Schindler, 1987 and in preparation). Likewise, warmer, drier weather causes increased reoxidation of reduced sulfur compounds found in wetlands, possibly increasing the probability of acid pulses to waters to which they drain.

In any case, it is important to act now to choose and secure sites for future ecosystem experiments, in areas where competing interests are still minimal, such as in the arctic; subarctic, in montane areas, and in wetlands. This fact has been recognized by the International Geosphere Biosphere Program (IGBP), and by the Royal Society of Canada's Global Change Committee, and it seems likely that ecosystem-scale experiments will play an important role in these programs.

7.11 SUMMARY

After 20 years we are still learning about more effective ways to use whole-ecosystem experiments. We have scarcely begun to tap the potential of this approach. In addition to affording direct tests of pollutants or management strategies, experiments in whole ecosystems can strengthen monitoring approaches by allowing us to test what variables are important to monitor. Similarly, experimental ecosystems allow us to calibrate smaller-scale approaches to detecting ecosystem stresses, so that we can be more certain of our interpretations of their results.

ACKNOWLEDGEMENTS

Gordon Koshinsky, Suzanne Bayley, and R.E. Hecky provided useful discussions of this topic and criticism of this manuscript. Work was supported by the Canadian Department of Fisheries and Oceans.

REFERENCES

Bayley, S.E. and Schindler, D.W. (1987) Sources of alkalinity in Precambrian Shield watershed streams under natural conditions and after stress (fire or acidification). In Hutchinson, T.C. and Meema K.M. (Eds), *Effects of Acidic Deposition on Forests, Wetlands and Agricultural Ecosystems*, Springer-Verlag, New York, pp. 531–40.

Bayley, S.E., Vitt, D.H., Newbury, R.W., Beaty, K.G., Behr, R. and Miller, C. (1987) Experimental acidification of a *Sphagnum*-dominated peatland: First year results. *Can. J. Fish. Aquat. Sci.* **44** (Suppl. 1), 194–205.

Brundin, L. (1956) Die bodenfaunistischen Seetypen und ihre Anwendbarkeit auf die Sudhabkugel. Zugleich eine Theorie der produktion-biologischen Bedeutung der glazialen Erosion. *Rep. Inst. Fresh-water Res. Drottningholm* **37**, 186–235.

Cook, R.B. and Schindler, D.W. (1983) The biogeochemistry of sulfur in an experimentally acidified lake. In: Halbert, R.O. (Ed.), *Ecol. Bull. (Stockholm)* **35**, 115–27.

Cook, R.B., Kelly, C.A., Schindler, D.W. and Turner, M.A. (1986) Mechanisms of hydrogen ion neutralization in an experimentally acidified lake. *Limnol. Oceanogr.* **31**, 134–48.

Davidson, G.A. (1984) Paleolimnological reconstruction of the acidification history of Lake 223 (ELA). MSc thesis, University of Manitoba, Winnipeg, MB.

Davies, I.J. (1989) Population collapse of the crayfish *Orconeates virilis* in response to experimental whole-lake acidification. *Can. J. Fish. Aquat. Sci.* **46**, 910–22.

Dillon, P.J., Reid, R.A. and Girard, R. (1986) Changes in the chemistry of lakes near Sudbury, Ontario following reductions in SO_2 emissions. *Water Air Soil Pollut.* **31**, 59–65.

Effler, S.W., Schafran, G.C. and Driscoll, C.T. (1985) Partitioning light attenuation in an acidic lake. *Can. J. Fish. Aquat. Sci.* **42**, 1707–11.

Eloranta, P.V. (1983) Physical and chemical properties of pond waters receiving warm-water effluent from a thermal power plant. *Water Res.* **17**, 133–40.

Findlay, D.L. and Kasian, S.E.M. (1987) Phytoplankton community responses to nutrient addition in Lake 226, Experimental Lakes Area, northwestern Ontario. *Can. J. Fish. Aquat. Sci.* **44** (Suppl. 1), 35–46.

France, R.L. (1987) Reproductive impairment of the crayfish *Orconectes virilis* in response to acidification of Lake 223. *Can. J. Fish. Aquat. Sci.* **44** (Suppl. 1), 97–106.

Gunn, J.M. and Keller, W. (1990) Biological recovery in an acid lake after reductions in the industrial emissions of sulfur. *Nature* **345**, 431–33.

Hecky, R.E. and Kilham, P. (1988) Nutrient limitation of phytoplankton in freshwater and marine environments: A review of recent evidence on the effects of enrichment. *Limnol. Oceanogr.* **33**, 796–822.

Hesslein, R. and Quay, P. (1973) Vertical eddy diffusion studies in the thermocline of a small stratified lake. *J. Fish. Res. Board Can.* **30**, 1491–500.

Holling, C.S. (1985) Resilience of ecosystems: Local surprise and global change. In: Malone, T.F. and Roederer, J.G. (Eds), *Global Change*, Cambridge University Press, Cambridge, MA, pp. 228–69.

Hutchinson, T.C. and Havas, M. (1986) Recovery of previously acidified lakes near Coniston, Canada following reductions in atmospheric sulphur and metal emissions. *Water Air Soil Pollut.* **28**, 319–333.

Kasian, S.E.M. and Campbell, P. A time series analysis of changes in light attenuation in experimentally acidified Lake 223 compared with natural variation in reference lakes. In review.

Keller, W. and Pitblado, J.R. (1986) Water quality changes in Sudbury area lakes: a comparison of synoptic surveys in 1974–1976 and 1981–1983. *Water Air Soil Pollut.* **29**, 285–96.

Kelly, C.A. and Rudd, J.W.M. (1984) Epilimnetic sulfate reduction and its relationship to lake acidification. *Biogeochemistry*. **1**, 63–77.

Kelly, C.A., Rudd, J.W.M., Hesslein, R.H., Schindler, D.W., Dillon, P.J., Driscoll, C., Gherini, S.A. and Hecky, R.E. (1987) Prediction of biological and neutralization in acid-sensitive lakes. *Biogeochemistry* **3**, 129–40.

Kelly, C.A., Rudd, J.W.M. and Schindler, D.W. (1990) Lake acidification by nitric acid: future considerations. *Water Air Soil Pollut.* **50**, 49–61.

Levine, S.N., Stainton, M.P. and Schindler, D.W. (1986) A radiotracer study of phosphorus cycling in a eutrophic Canadian Shield lake, Lake 227, northwestern Ontario. *Can. J. Fish. Aquat. Sci.* **43**, 366–78.

Levine, S.N. (1983) Natural mechanisms that ameliorate nitrogen shortages in lakes. PhD thesis, University of Manitoba, Winnipeg, MB.

Mills, K.H. and Chalanchuk, S.M. (1987) Population dynamics of lake whitefish (*Coregonus clupeaformis*) during and after the fertilization of Lake 226, the Experimental Lakes Area. *Can. J. Fish. Aquat. Sci.* **44** (Suppl. 1), 55–63.

Mills, K.H., Chalanchuk, S.M., Mohr, L.C. and Davies, I.J. (1987) Responses of fish populations in Lake 223 to 8 years of experimental acidification. *Can. J. Fish. Aquat. Sci.* **44** (Suppl. 1), 114–25.

Müller, P. (1980) Effects of artificial acidification on the growth of epilithiphyton. *Can. J. Fish. Aquat. Sci.* **37**, 355–63.

Nicolson, J.A. (1975) Water quality and clear cutting in a boreal forest ecosystem. *Proceedings of Canadian Hydrology Symposium 75*. NRCC No. 15195, 11–14 August, Winnipeg, MB, pp. 734–8.

Patalas, K. (1970) Primary and secondary production in lake heated by thermal power plant. *Proc. 1970 Ann. Tech. Meeting, Inst. of Environ. Sciences*, April, Boston, MA, pp. 267–71.

Playle, R.C. (1985) The effects of aluminum on aquatic organisms: (1) alum additions to a small lake, and (2) aluminum-26 tracer experiments with minnows. M.Sc. thesis, University of Manitoba, Winnipeg, MB.

Quay, P.D. (1977) An experimental study of turbulent diffusion in lakes. PhD thesis, Columbia University, New York.

Rudd, J.W.M., Kelly, C.A., St. Louis, V., Hesslein, R.H., Furutani, A. and Holoka, M. (1986) Microbial consumption of nitric and sulfuric acids in acidified north temperate lakes. *Limnol. Oceanogr.* **31**, 1267–80.

Rudd, J.W.M., Kelly, C.A., Schindler, D.W. and Turner, M.A. (1988) Disruption of the nitrogen cycle in acidified lakes. *Science (Wash., DC)* **240**, 1515–17.

Rudd, J.W.M., Kelly, C.A., Schindler, D.W. and Turner, M.A. (1990) A comparison of the acidification efficiency of nitric and sulfuric acids by two whole lake addition experiments. *Limnol. Oceanogr.* **35**, 663–679.

Santschi, P.H., Nyfeller, U.P., Anderson, R.F., Schiff, S.L., O'Hara, P. and Hesslein, R.H. (1986) Response of radioactive trace metals to acid–base titrations in controlled experimental ecosystems: comparison of results from enclosure and whole-lake radio-tracer additions. *Can. J. Fish. Aquat. Sci.* **43**, 60–77.

Schiff, S.L. and Anderson, R.F. (1987) Limnocorral studies of chemical and biological acid neutralization in two freshwater lakes. *Can. J. Fish. Aquat. Sci.* **44** (Suppl. 1), 173–87.

Schindler, D.W. (1971) Carbon, nitrogen, phosphorus and the eutrophication of freshwater lakes. *J. Phycol.* **7**, 321–9.

Schindler, D.W. (1974) Eutrophication and recovery in experimental lakes: Implications for lake management. *Science (Wash., DC)* **184**, 897–9.

Schindler, D.W. (1977) Evolution of phosphorus limitation in lakes: natural mechanisms compensate for deficiencies of nitrogen and carbon in eutrophied lakes. *Science (Wash., DC)* **195**, 260–2.

Schindler, D.W. (1980) Experimental acidification of a whole lake: A test of the oligotrophication hypothesis. In: Drablos, D. and Tollan, A. (Eds), *Ecological Impact of Acid Precipitation*. Proceedings of an International Conference, Sandefjord, Norway, 11–14 March, SNSF Project, Oslo, pp. 370–4.

Schindler, D.W. (1986) The significance of in-lake production of alkalinity. *Water Air Soil Pollut.* **30**, 931–44.

Schindler, D.W. (1987) Recovery of Canadian lakes from acidification. In: Barth, H. (Ed.), *Reversibility of Acidification*, Elsevier, London, pp. 2–13.

Schindler, D.W. (1988a) Experimental studies of chemical stressors on whole lake ecosystems. *Verh. Internat. Verein. Limnol.* **23**, 11–41.

Schindler, D.W. (1988b) Effects of acid rain on freshwater ecosystems. *Science (Wash., DC)* **239**, 149–57.

Schindler, D.W. (1990) Experimental perturbations of whole lakes as tests of ecosystem structure and function. *Oikos* **57**, 25–41.

Schindler, D.W., Bayley, S.E. Curtis, P.J., Parker, B.J., Stainton, M.P. and Kelly, C.A. (1991) Natural and experimental factors affecting the abundance and cycling of dissolved organic substances in Precambrian Shield lakes and their watersheds. *Hydrobiologia* (In press).

Schindler, D.W., Brunskill, G.J., Emerson, S., Broecker, W.S. and Peng, T.-H. (1972) Atmospheric carbon dioxide: Its role in maintaining phytoplankton standing crops. *Science (Wash., DC)* **177**, 1192–4.

Schindler, D.W., Kling, H., Schmidt, R.V., Prokopowich, J., Frost, V. E., Reid, R.A. and Capel, M. (1973) Eutrophication of Lake 227 by addition of phosphate and nitrate. Part 2. The second, third and fourth years of enrichment, 1970, 1971 and 1972. *J. Fish. Res. Board Can.* **30**, 1415–40.

Schindler, D.W., Hesslein, R.H. and Kipphut, G. (1977) Interactions between sediments and overlying waters in an experimentally-eutrophied Precambrian Shield Lake. In: Golterman, H.L. (Ed.), *Interactions Between Sediments and Freshwater*. Proceedings of Symposium, Amsterdam, September 1976, W. Junk, The Hague, PUDOC, Wageningen, pp. 235–43.

Schindler, D.W., Newbury, R.W., Beaty, K.G., Prokopowich, J., Ruszczynski, T. and Dalton, J.A. (1980) Effects of a windstorm and forest fire on chemical losses from forested watersheds and on the quality of receiving streams. *Can. J. Fish. Aquat. Sci.* **37**, 328–34.

Schindler, D.W., Mills, K.H., Malley, D.F., Findlay, D.L., Shearer, J.A., Davies, I.J., Turner, M.A., Linsey, G.A. and Cruickshank, D.R. (1985) Long-term ecosystem stress: The effects of years of acidification on a small lake. *Science (Wash., DC)* **228**, 1395–401.

Schindler, D.W., Hesslein, R.H. and Turner, M.A. (1987) Exchange of nutrients between sediments and water after 15 years of experimental eutrophication. *Can. J. Fish. Aquat. Sci.* **44** (Suppl. 1), 26–33.

Schindler, D.W., Beaty, K., Fee, E.J., Cruickshank, D., De Bruyn, E., Findlay, D., Linsey, G., Shearer, J.A., Stainton, M.P. and Turner M.A. (1990) Effects of climatic warming on lakes of the central boreal forest. *Science (Wash., DC)* **250**, 967–970.

Shearer, J.A., Fee, E.J., DeBruyn, E.R. and DeClercq, D.R. (1987) Phytoplankton productivity changes in a small, double-basin lake in response to termination of experimental fertilization. *Can. J. Fish. Aquat. Sci.* **44** (Suppl. 1), 47–54.

Warwick, W.W. (1980) Palaeolimnology of the Bay of Quinte, Lake Ontario: 2800 years of cultural influence. *Can. Bull. Fish. Aquat. Sci.* **206**, 1–117.

Wright, R.F., Lotse, E. and Semb, A. (1988) Reversibility of acidification shown by whole-catchment experiments. *Nature* **334**, 670–5.

8 The Role of Fire in Determining Stream Water Chemistry in Northern Coniferous Forests

S.E. BAYLEY and D.W. SCHINDLER
Department of Botany, University of Alberta, Edmonton, Alberta T6G 2E9, Canada

8.1 INTRODUCTION

Fire has an important role in structuring ecosystems. Knowledge of the effects of fire on the ecosystems allows proper management of those ecosystems in terms of regeneration and productivity of vegetation, nutrient retention, and land/water interactions. Fires can be viewed as large-scale ecosystem experiments, but because forest fires usually occur without warning there are seldom pre-fire data, and effects are poorly understood. This is particularly true of the effect of fire on nutrient losses from watersheds. Most of the published studies were started one or two years after the fire (Wright, 1976; Nakane *et al.*, 1983).

Several papers have stated general hypotheses concerning the role of disturbance (usually clearcutting) in determining chemical losses from forested watersheds via streamflow (Likens *et al.*, 1970, 1977; Johnson and Swank, 1973; Vitousek *et al.*, 1979; Swank, 1988). There have been far fewer papers hypothesizing the effects of wildfire on stream chemistry (Tiedemann *et al.*, 1978; Wright, 1976). Few of these hypotheses have been tested, for seldom have watersheds been burned after they have been studied for a reference period long enough to allow conclusive examination of the effects of wildfire. Here, we review the results of other stream studies from burned watersheds and examine a few hypotheses using a 17-year data set of stream chemistry from two burned boreal forest watersheds at the Experimental Lakes Area (ELA) in northwestern Ontario, Canada. The main fire studies reviewed here are: a wildfire in mixed coniferous–deciduous forest in northern Minnesota (Little Sioux fire) (Wright, 1976); a wildfire in Ponderosa pine/Douglas fir forest in the mountains of north-central Washington at Entiat Experimental Forest (Tiedemann *et al.*, 1978); a wildfire in a red pine forest in the mountains of Etajima Island, Japan (Nakane *et al.*, 1983); a clearcut and slash–burn study in a western hemlock/western red cedar/

Ecosystem Experiments. Edited by H.A. Mooney *et al.*

Douglas fir forest in British Columbia (Haney fire—Feller and Kimmins, 1984).

8.2 THE ALKALINE ASH HYPOTHESIS

Most authors assume that aggrading forests will accumulate alkalinity that would otherwise be flushed into streams and lakes. This hypothesis seems to be rooted in the general knowledge that growing trees take up base cations (Nilsson *et al.*, 1982; Marks and Bormann, 1972) and give up H^+ to the soil (Brinkley and Richter, 1987). Rosenqvist (1978) and Krug and Frink (1983, 1984) believed that this differential biological uptake would acidify soils, causing the acidification of water percolating through soils in aggrading forests. They hypothesized that most forest soils of the eastern USA and Scandinavia are acidifying because they are recovering from earlier land disturbance. It is implied that natural acidification, not acid precipitation, is causing acidification of European and North American soils, streams and lakes. Neither of the above authors have shown a relationship between the acidification of soils and surface waters. Brown (1984) carried this hypothesis to its logical conclusion, suggesting that fire-suppression policies in the twentieth century have caused the acidification of lakes and streams because the natural fires that periodically returned alkalinity to soils in previous centuries are being prevented.

There are numerous studies that document an increase in base cation concentrations in soils after fire (Grier, 1975; Grier and Cole, 1971; Smith, 1970; McColl and Grigal, 1977; Woodmansee and Wallach, 1981), some documentation of increases in base cations in streams (see below) and very few data on release of acid anions other than nitrate. Thus the widely held view is that fires in watersheds cause an alkaline ash (or base cation release) that increases the alkalinity and pH of streams draining the watersheds.

8.3 THE NITRATE HYPOTHESIS

A second widely held view is that ecosystems disturbed by clearcutting, burning, or treatment with herbicide lose large amounts of nitrate in stream flow (Likens *et al.*, 1969, 1970; Tiedemann *et al.*, 1978; Feller and Kimmins, 1984; Brown *et al.*, 1973). Although Vitousek *et al.* (1979) showed that some forest ecosystems do not produce a pulse of nitrate after disturbance, most forest ecosystems do. He suggested the more fertile the ecosystem, the greater the loss of nitrate. Nitrification is generally enhanced after fire. Woodmansee and Wallach (1981) state that the mineralization to NH_4^+ is followed by nitrification and denitrification, particularly if the trees are killed but the N-containing organic residues remain. In an earlier paper on the watersheds at the Experimental Lakes Area (ELA), Schindler *et al*. (1980) found that nitrogen losses after fire were elevated, but only by 2–3-fold, and only for a few years. The watersheds analysed in that study have now burned a second time, causing extreme damage to the vegetation. The reference watershed studied by Schindler *et al*. also burned in 1980, after nine years of data had been collected on the mature forest stream. The period of record included years with both below- and above-average moisture conditions.

8.4 THE EVAPOTRANSPIRATION HYPOTHESIS

A third hypothesis we will examine is that burned watersheds typically export more water than unburned watersheds. The increase in runoff is due to the reduction in evapotranspiration from burned watersheds. Most studies have documented an increase in water yield after disturbance (see review by Bosch and Hewlett, 1982); however, the fast-growing mountain ash of Australia apparently increases evapotranspiration and consequently lowers runoff after fire (Attiwill, this volume). Several Canadian studies also did not detect higher runoff after fire or logging, probably due to the limited soil water storage capacity and thin soils in many boreal forest areas of Canada (Hetherington, 1987) or to year-to-year differences in precipitation and evapotranspiration.

The effect of fire on hydrological and biogeochemical cycles can vary depending on the ecosystem, the frequency and intensity of the fire, the season the fire occurs, and the meteorological conditions after the fire (Woodmansee and Wallach, 1981).

We will examine the above hypotheses comparing published studies of wildfire and one slash burning study with our long-term data set (17 years) from two basins in the Experimental Lakes Area of northwestern Ontario, Canada. We compare stream chemistry and hydrology before and after forest fires in two terrestrial watersheds to determine the effect of forest fire on stream chemistry and hydrology. We are able to compare basins that are side-by-side, as is typical of most fire studies (Wright, 1976), as well as before-and-after comparisons within the same basin. In addition, we compare single and double burns within the same basin. We have eight years of data after the most recent fire, so that we are able to document the initial recovery of the stream chemistry and hydrology.

8.5 AN EXPERIMENTAL TEST

8.5.1 DESCRIPTION OF THE AREA

The Experimental Lakes Area is located in northwestern Ontario 300 km east of Winnipeg on the Precambrian Shield. The controlled area consists of 46

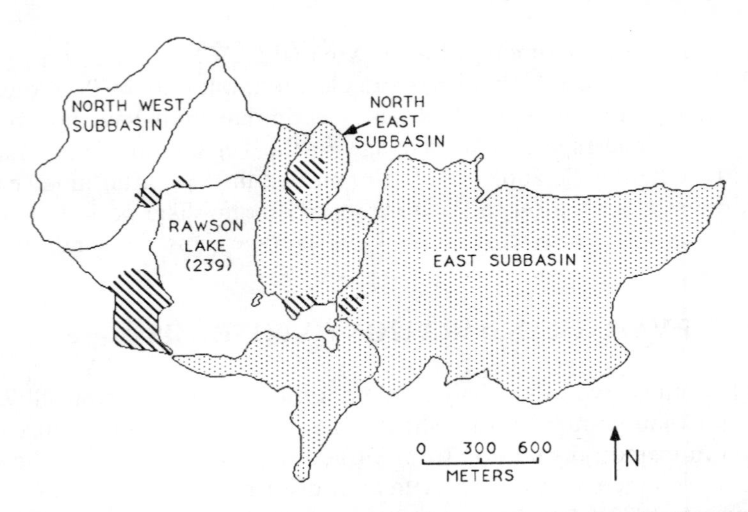

Figure 8.1 Map of the Rawson Lake watershed (L239) showing the Northwest and East Basins. Grey areas burned in 1974. All of the watershed but the hatched areas burned in 1980.

small lakes and their watersheds. The two small basins analyzed here are in the drainage of Lake 239, which has been monitored since 1968 (Fig. 8.1). Samples for chemical analysis have been collected weekly during the ice-free season since 1970 in the Northwest (NW) Basin, and since 1971 in the East (E) Basin. The period analyzed here is for 1971–87, during which sampling and analytical methods were similar, and any changes in methods were carefully intercalibrated. The NW and E Basins drain upland areas (56 ha and 170 ha, respectively) and have an average slope of approximately 10 degrees (Table 8.1).

Table 8.1 Physical characteristics of the Lake 239 subbasins

Characteristic	Basin: Northwest	Basin: East
Area (ha)	56.4	170.3
Average slope (%)	11	10
Type of cover (%):		
Rock outcrop	21	10
Wetland	3	4
Upland with shallow mineral soil	72	80
Approximate depth of overburden (m) in the bottom of the basin	3	9

In 1973 a severe windstorm caused widespread damage to the forests in the E Basin. Blowdown was up to 100% on ridgetops, and averaged about 80%. This was followed by a wildfire in late June of 1974 which burned the E Basin. The NW Basin was not burned in 1974 and was little affected by the windstorm. A more detailed description of the fire and windstorm, and the effects of the fire and windstorm on yields of N, P and K for the years following the fire, were reported in Schindler *et al.* (1980).

In June 1980 a fire reburned the E Basin. The NW Basin burned for the first time in 1980. For convenience in analysis we have divided the data into three blocks. For the NW Basin, the pre-fire block includes 1971–80 (until the fire), the years immediately following the fire, when severe changes in the chemistry and runoff from the basin were expected (1980–82), and the recovery period (1983–87). The corresponding periods for the E Basin were: pre-fire, 1971–73; post-fire 1, 1974–76; recovery 1, 1978–79; post-fire 2, 1980–87.

Before fire, the terrestrial ecosystems were dominated by a mature (approx. 100-year-old) forest of *Pinus banksiana* (jackpine) with some *Picea mariana* (black spruce), *Betula papyrifera* (paper birch), and *Populus tremuloides* (trembling aspen).

After the first fire the terrestrial areas re-vegetated rapidly with small jackpine, black spruce, aspen, birch, black alder (*Alnus nigra*), balsam poplar (*Populus balsamifera*), and pincherry (*Prunus pennsylvanica*). Shrub and herbaceous vegetation were dense for a few years following the fire: fireweed (*Epilobium angustifolium*), black bindweed (*Polygonum convolutus*), raspberry (*Rubus strigosus* Michx.), and blueberry (*Vaccinium* spp.). After the second burn, similar species were observed, except that recovery was much slower and conifers were not as numerous.

Both streams were gauged with V-notch weirs or flumes with recorders, providing a continuous flow record during the ice-free season (Beaty, 1981). Samples for chemical analyses were collected weekly, except during spring melt when sampling was more frequent. Chemical analysis for perishable forms (pH, DIC, alkalinity, NO_3-N, NH_4-N) were analyzed the same day at the ELA field camp, while the less perishable ions (K^+, Ca^{2+}, Mg^{2+}, Na^+, SO_4^{2-} and Cl^-) were analyzed at the Freshwater Institute in Winnipeg, Manitoba.

The methods of Stainton *et al.* (1977) were used for all ions except for SO_4^{2-}. The SO_4^{2-} was initially measured (1971 and 1972) using the gravimetric technique with barium chloride (APHA, 1967) and then with the ion exchange technique (Stainton *et al.*, 1977). These methods are now known to measure sulfate plus organic anions (Kerekes *et al.*, 1984). Starting in 1980 Dionex ion chromatography was used. The ion exchange method and the ion chromatography technique were intercalibrated by analyzing all samples with both methods for two years and comparing the results. Regressions of ion

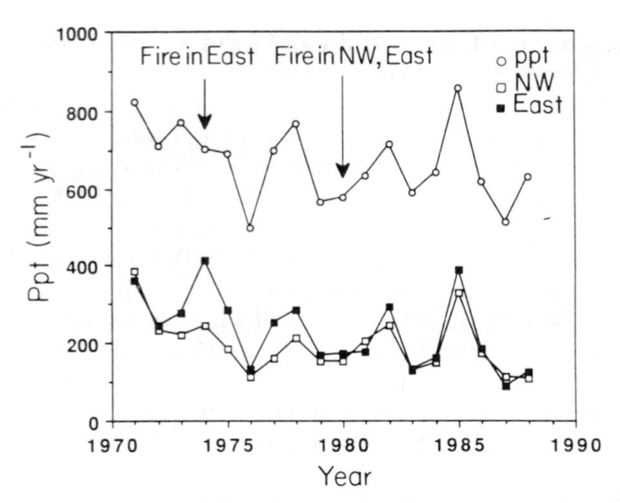

Figure 8.2 Annual precipitation and stream discharge from the Northwest and East Basins from 1971 to 1987.

exchange vs chromatography SO_4^{2-} show good agreement ($r^2 = 0.81$) for the NW Basin and the E Basin ($r^2 = 0.71$).

In this paper we will present the annual data as concentration because most of the comparable literature is as concentration. Mass export per hectare will be presented in a later paper. All concentration data are presented as volume weighted mean annual concentrations based on a water year from 1 November through 31 October. In the fire years of 1974 and 1980 the spring runoff precedes the fires and hence the mean annual concentrations minimizes the effect of the fires by incorporating both burned and unburned components of the annual concentration.

8.5.2 EFFECT OF FIRE ON WATER EXPORT

Fire in the E Basin in 1974 caused an increase in water export compared to the unburned NW Basin by 69% in 1974 and by 55% in 1975 (Fig. 8.2). Yield of water remained significantly higher ($p = 0.05$) in the burned basin compared to the unburned NW basin for six years after the fire. Precipitation on the watersheds averaged 680 mm/year, with runoff from the E Basin averaging 42% of precipitation in the three years after the fire. The runoff in the NW Basin averaged 30% of precipitation, both before and after the 1980 fire.

Two other studies of wildfires in pine forests (the Little Sioux fire by Wright (1976) and the Etajima Island fire in Japan by Nakane *et al.* (1983)) showed that runoff from burned watersheds was greater than from the adjacent unburned watersheds.

8.5.3 EFFECT OF FIRE ON NUTRIENTS

Nutrient export from the ELA watersheds increased dramatically after fires. Concentrations of total nitrogen in the NW Basin increased by 68% in the first two years after fire (Fig. 8.3), while the E Basin increased by 92% in the first two years after fire and has remained high (Fig. 8.3). The additional effect of the second burn in the E Basin was negligible. The total N concentrations in the E Basin have not returned to the pre-fire (1971–73) levels. In contrast, the NW Basin shows different results with the average total N concentration returning rapidly to pre-fire levels within two years after the first fire (Fig. 8.3). Total N concentrations in the NW stream increased in other years that are unrelated to the effect of fire. For example, in the wet year of 1977 concentrations of total N in the NW Basin also increased dramatically prior to any fires in that basin. This may be due to the higher dissolved organic N in response to the higher precipitation in that year.

Phosphorus concentrations (TDP and TP) also appeared to increase after the first burn in both basins (Figs 8.4A and 8.4B). In the E Basin the fire effect on the stream TP concentrations was greater than its effect in the NW stream. In the NW Basin the highest annual average total P occurred in 1971, before any watershed disturbance. The next highest concentrations occurred two to six years after the burn. In the NW Basin the years of higher than average

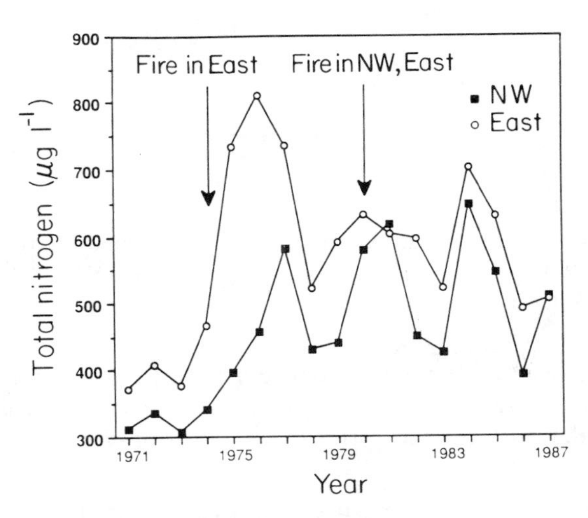

Figure 8.3 Annual total nitrogen concentration from 1971 to 1987 in Northwest and East streams.

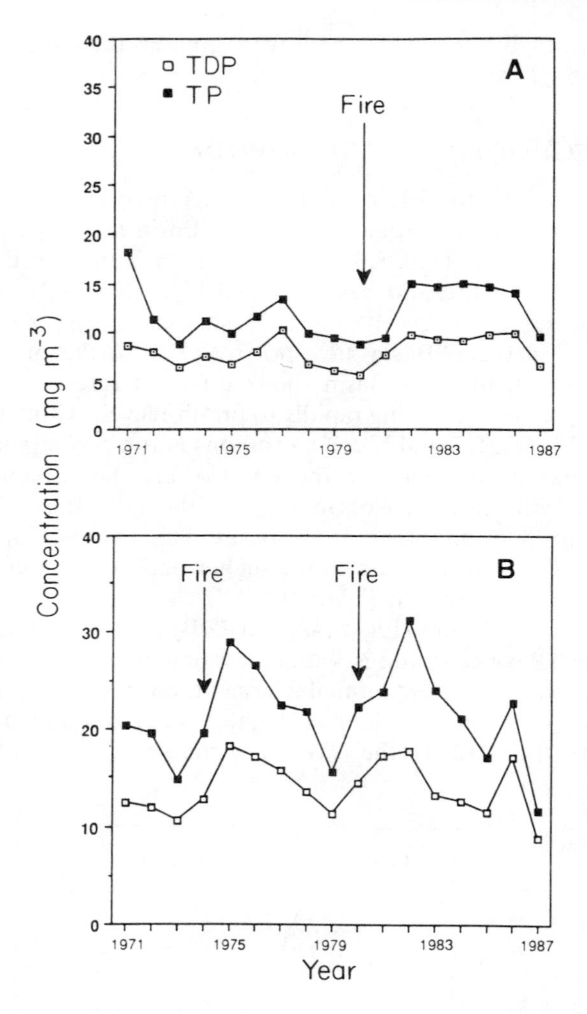

Figure 8.4 Annual concentrations of total dissolved phosphorus and total phosphorus from 1971 to 1987. A: Northwest stream; B: East stream.

precipitation also had the highest concentrations of total P and TDP prior to the fire, although regressions of stream flow versus concentration showed no significant relationship either prior to or following fire.

8.5.4 EFFECT OF FIRE ON BASE CATION RELEASE

In contrast to the soils studies whereby base cation concentrations increase after fire (Grier, 1975; Smith, 1970), studies documenting the export of base cations in streams after wildfire show a variety of responses. Nakane *et al.* (1983) (Etajima fire) detected increased concentrations of K^+, no difference in Na^+ and Mg^{2+} concentrations and decreased concentrations of Ca^{2+} after fire. Wright (1976) did not detect a significant increase in Ca^{2+}, Mg^{2+} or Na^+ concentrations in the Little Sioux fire. However, he detected an increase in K^+ export after fire. Two other studies of fire also show that base cation increases are not routinely detected in streamwater after fire. The Entiat fire in eastern Washington (Tiedemann *et al.*, 1978) showed decreased Ca^{2+} concentrations after fire and a slight increase in K^+, Mg^{2+}, and Na^+ concentrations immediately after fire. However, this increase was obscured by an increase in water flow that diluted the base cation concentrations. The Haney, British Columbia clearcut and slash burning study by Feller and Kimmins (1984) showed increased concentrations of K^+ and Mg^{2+} after burning but detected a greater base cation response to clearcutting alone than to clear-cutting plus slash burning. Increased concentrations of the base cations Ca^{2+}, Mg^{2+}, and Na^+ after wildfire appear to be more variable than increased concentrations of K^+.

In our study, concentrations of some base cations increased after fire in streams from both the NW and E Basins (Figs 8.5A and 8.5B). After the first burn in both basins, increased concentrations of K^+, Ca^{2+}, and Mg^+ were observed. Sodium concentrations were elevated in the NW Basin but not in the E Basin after the first burn. Burning the E Basin a second time caused an increase in concentration in Ca^{2+}, Mg^{2+}, and Na^+ but not in K^+ (Fig. 8.5B).

Analysis of base cations on a monthly basis shows that concentrations have remained higher than pre-fire levels even in the 1983–87 recovery period. Stream concentrations of Ca^{2+} and Mg^{2+} (Figs 8.6A and 8.6B) and of Na^+ and K^+ (Figs 8.7A and 8.7B) were higher after the fire than before through-out the year. Seasonal changes in concentrations of K^+ were detected in the NW Basin, with higher levels in the spring and fall than in the summer, as is expected for this nutrient (Fig. 8.7B). The K^+ concentrations were low before the fire and significantly higher immediately after the fire. They have declined approximately 50% since the highest levels after the fire, but are still above pre-fire values (Fig. 8.7B). Although rapidly growing trees are known to take up K and other base cations, even after eight years, concentrations in the stream are significantly higher than concentrations draining the mature forest before the fire (1971–80).

Comparison of the base cation release from the E Basin after the 1980 and 1974 fires shows higher concentrations after the 1980 fire. While this may be caused by dissolution of ash, breakdown of organic matter, or increased weathering, the increases may also be due to the diminished runoff after the 1980 fire. The differences between the smaller increases detected after the 1974 fire compared to the 1980 fire may be due to a dilution effect, since runoff was increased by 69% after the 1974 fire and runoff did not increase after the 1980 fire.

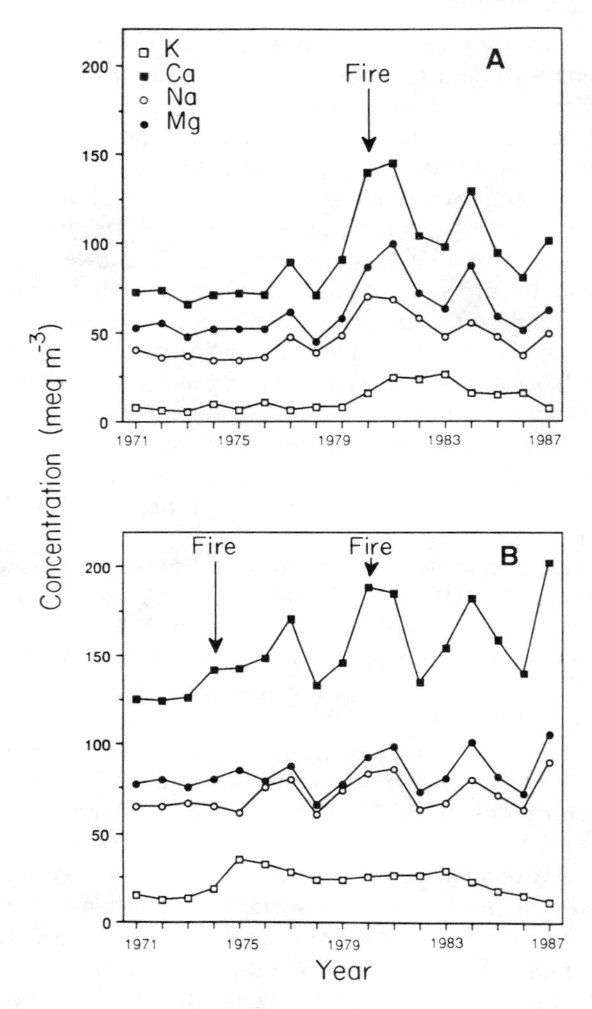

Figure 8.5 Annual concentrations of base cations from 1971 to 1987. A: Northwest stream; B: East stream.

8.5.5 EXPORT OF STRONG ACID ANIONS AFTER FIRE

Review of the stream studies on the effect of fire on sulfate and chloride concentrations in burned watersheds shows no data available for wildfire and few data available for controlled fires.

In our ELA watersheds, large amounts of sulfate and chloride were released as a pulse into the streams. Sulfate concentrations in the NW Basin increased over 350% after the fire, remaining elevated even six years after

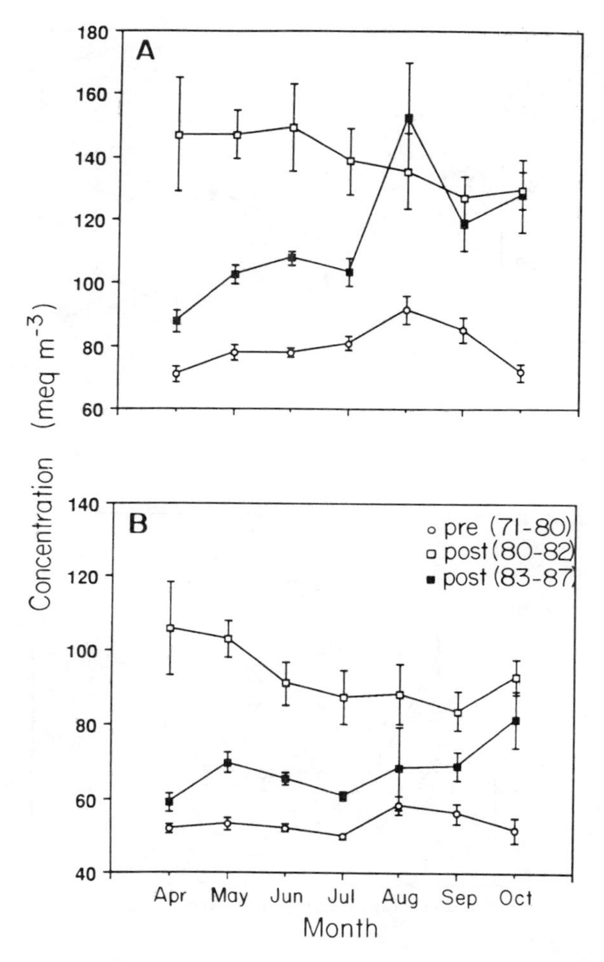

Figure 8.6 Pre-fire and post-fire mean monthly concentrations in the Northwest stream. A: Calcium; B: magnesium.

burning (Fig. 8.8A). Peak concentrations of SO_4^{2-} reached 634 meq m^{-3} in 1981. In the years since the fire there was slight suggestion of a seasonal pattern in the NW stream with higher SO_4^{2-} concentrations in the spring and fall than midsummer (Fig. 8.8B). There was also an increase in annual SO_4^{2-} concentrations in the E Basin after the 1974 and 1980 fires (Fig. 8.8A), although not as much as in the NW Basin. Unfortunately, the SO_4^{2-} analyses prior to the 1974 fire were measured with the gravimetric barium chloride technique, which is not comparable to either the ion exchange or ion chromatographic techniques used since 1974. Therefore the NW Basin has

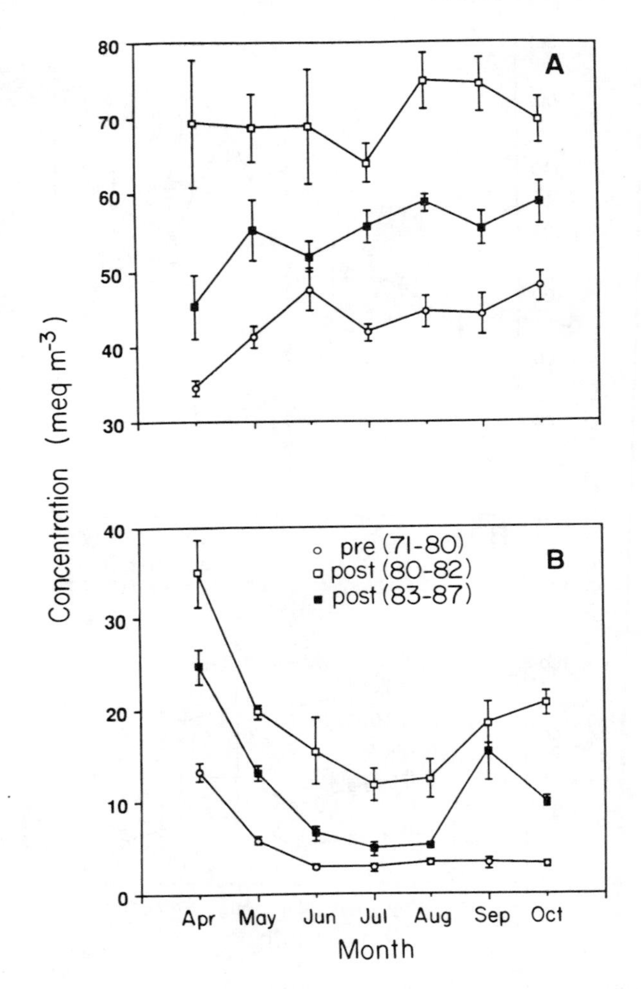

Figure 8.7 Pre-fire and post-fire mean monthly concentrations in the Northwest stream. A: Sodium; B: potassium.

the best data set to demonstrate the large increase in SO_4^{2-} concentrations after fire.

In the NW stream after fire (1980–81), there is a significant relationship between Ca^{2+} plus Mg^{2+} versus SO_4^{2-} concentrations ($r^2 = 0.75$) (Fig. 8.9A). This suggests that the base cation export in this stream was driven by the sulfate export. No such relationship existed for NO_3^-. Likens *et al.* (1970) detected a good relationship between nitrate export and base cation export after clearcutting (and herbicide) (Fig. 8.9B). While higher base cations

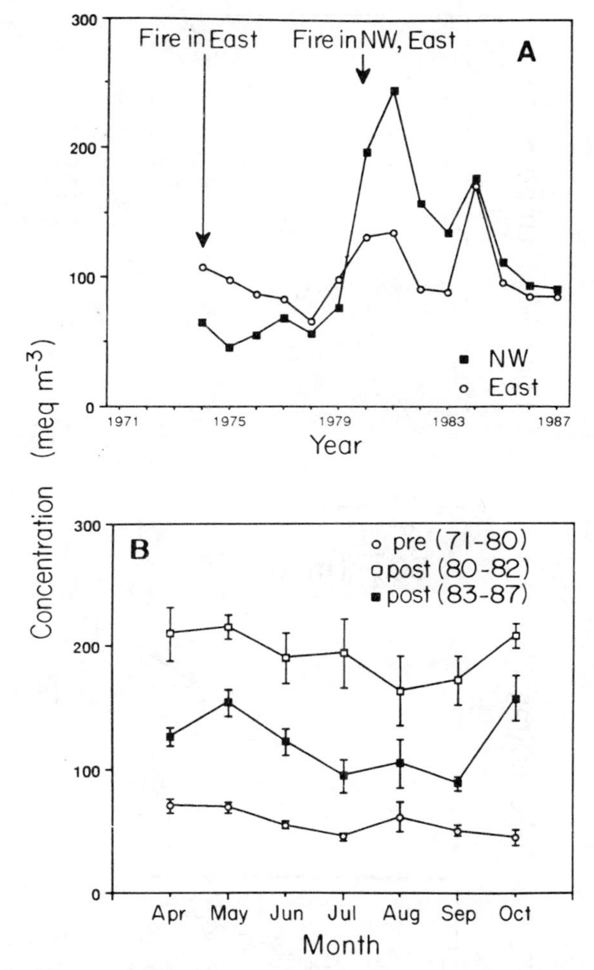

Figure 8.8 Sulfate concentrations in ELA streams. A: Mean annual concentrations in the Northwest and East streams from 1974 to 1987; B: mean monthly SO_4^{2-} concentrations in the Northwest stream before and after the 1980 fire.

concentrations were detected after disturbance at Hubbard Brook than after the ELA fire, SO_4^{2-} concentrations at Hubbard Brook declined (Table 8.2).

Chloride concentrations in stream water followed the same pattern as SO_4^{2-}, with increases of 300% after the fire in the NW Basin (Fig. 8.10A). Both the first and second burns in the E Basin produced sharp pulses of Cl^-. Chloride concentrations were higher in spring and fall and lower in summer just after the fire in the NW Basin. After three to seven years the concentration had declined to the point where concentrations of Cl^- were

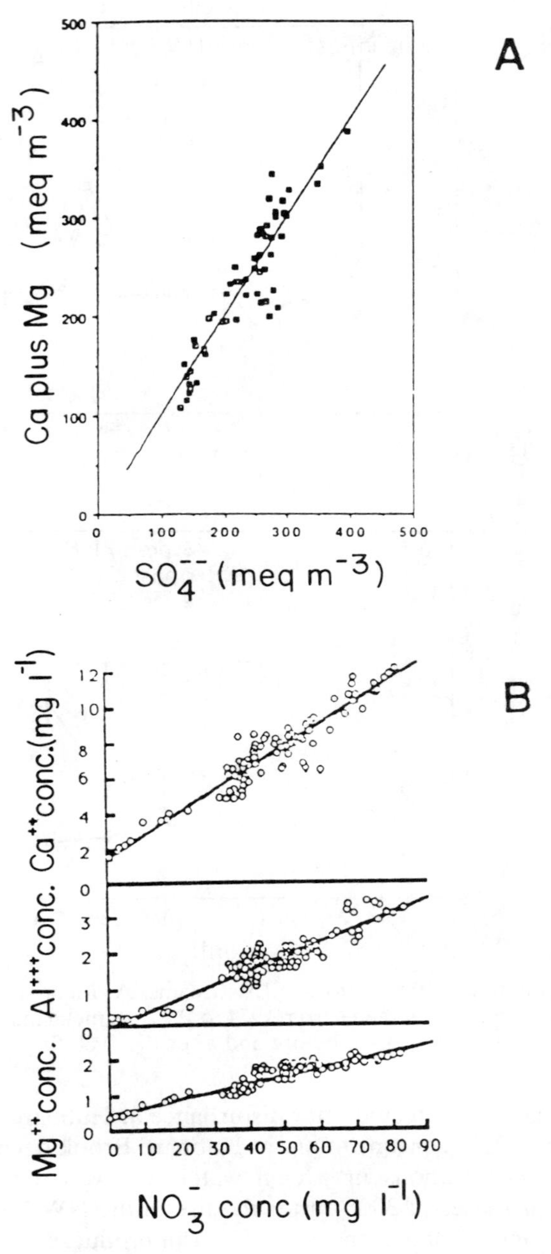

Figure 8.9 A: Regression of Ca^{2+} plus Mg^{2+} versus SO_4^{2-} in the Northwest stream after fire (1980–1981); B: regressions of Ca^{2+}, Al^{3+}, and Mg^{2+} versus NO_3^- concentrations in Hubbard Brook after disturbance (from Likens *et al.*, 1970).

Table 8.2 Nitrate versus sulfate control of base cation leaching after forest disturbance

	ELA NW watershed			Hubbard Brook no. 2		
	Pre-fire 1971–80	Fire year 1980	2nd year 1981	Undisturbed 1965–6	Cut 1966–67	2nd year after 1967–68
NO_3^--N	0.8	7.9	17.5	15	619	853
SO_4^{2-}	40	206	246	142	79	77
Ca^{2+}	83	139	145	90	322	377
Σ base cations	184	311	337	163	548	644

ELA Northwest basin pre-fire (mean 1971–1980), 1980 fire year, second year after fire (1981). Hubbard Brook watershed no. 2 (Likens *et al.*, 1970). Pre-logging 1965–66, first year post-disturbance (1966–67), second year post-disturbance (1967–68) (data in meq m^{-3}).

significantly lower than either the pre-fire or immediately post-fire (1980–82) values (Fig. 8.10B). The release of Cl^- after clearcutting and herbicide was also observed in the Hubbard Brook study as well (Likens *et al.*, 1970, 1977). Chloride concentrations increased only after clearcutting in Haney, BC, but not after clearcutting plus slash burning (Feller and Kimmins, 1984). These pulsed losses of Cl^- after disturbance suggest that chloride is not conservative in natural ecosystems in the short term, and should be used with great caution as a conservative indicator.

Nitrate is the third important acid anion that increases dramatically in concentration after fire. Concentrations in the ELA study site (NW stream) were approximately 0.8 meq m^{-3} prior to the fire, and averaged 17.5 meq m^{-3} in 1982, two years after the fire (Fig. 8.11A). Nitrate concentration declined rapidly in subsequent years, unlike the concentrations of sulfate (Fig. 8.8A). Nitrate exhibited a pronounced seasonal cycle in the two years after the fire with higher concentrations during the spring melt (Fig. 8.11B). The growing biomass presumably assimilated the nitrate during the summer even after the fire.

Concentrations of NO_3^- released in the ELA watersheds, even after fire, were 10-fold lower than those released from undisturbed Hubbard Brook streams (Likens et al., 1977). Nitrate concentrations increased after the Entiat fire (Tiedemann *et al.*, 1978) from 16 to 560 meq m^{-3}, considerably higher than our fire response. In the Little Sioux fire, Wright (1976) did not detect any increase in nitrate release, probably because the study did not commence until the second year after the fire. However, Wright's mixed southern boreal forest ecosystem is more similar to the ELA watersheds than to any of the other study forests. The second burn (1980) in the E Basin still elicited a large release of nitrate (Fig. 8.11A), although the response was lower than after the first burn (1974).

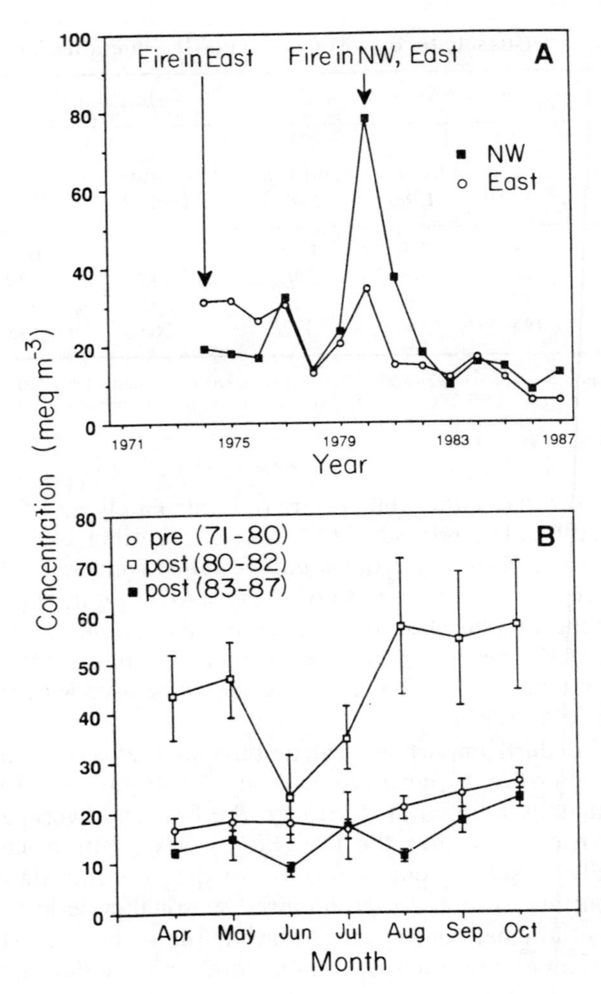

Figure 8.10 Chloride concentrations in ELA streams. A: Mean annual concentrations in the Northwest and East stream from 1971 to 1987; B: mean monthly Cl^- concentrations in Northwest stream before and after the 1980 fire.

Vitousek *et al.* (1979) suggested that most forest ecosystems release nitrate into soil water after disturbance. They suggested that some forest ecosystems, particularly less fertile sites, may bind NO_3^- and NH_4^+ initially and then eventually release NO_3^- in runoff. Richer sites release NO_3^- rapidly from soils. Tiedeman *et al.* (1978) found increased NO_3^- released after fire, which they speculated was initially caused by decreased plant uptake and

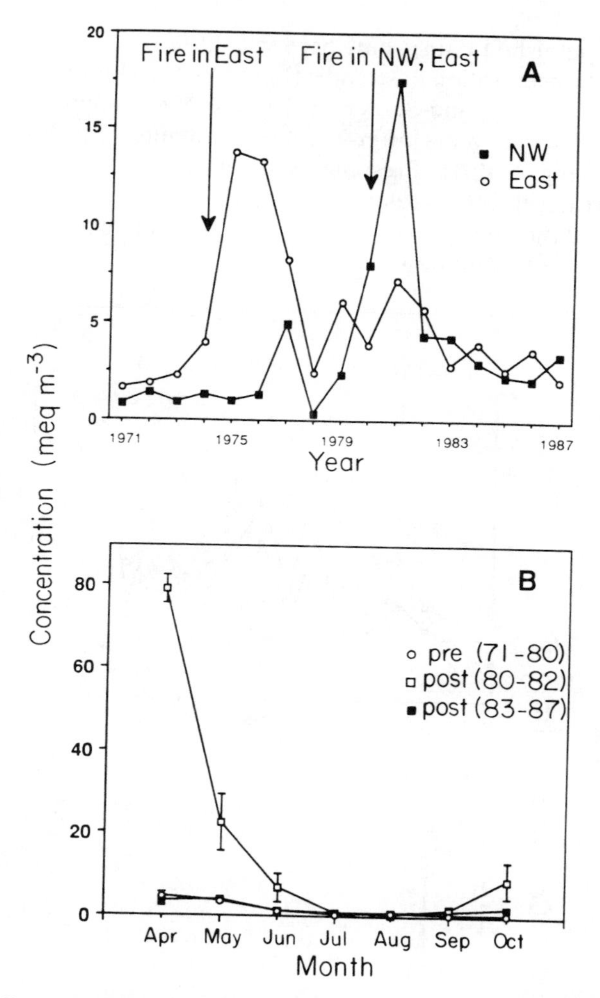

Figure 8.11 Nitrate–nitrogen concentrations in ELA streams. A: Mean annual concentrations in the Northwest and East streams from 1971 to 1987; B: mean monthly NO_3^--N concentrations in Northwest stream before and after the 1980 fire.

subsequently caused by nitrification. Likens (1969) found that clearcut Hubbard Brook watersheds released very high concentrations of NO_3^-, some of which they attributed to nitrification. Likens (1969) did not detect an increase in NH_4^+ in stream runoff after disturbance, although there was a large increase in soil ammonium. The 1974 fire disturbance in the ELA watersheds produced increased soil concentrations of NH_4^+ and NO_3^-

(Schindler, unpublished data), but there was little or no increase in annual average NH_4^+ concentration in streams (Fig. 8.12A). Monthly concentrations of NH_4^+ (in June, July, and September) in the NW stream increased in the period three to seven years after the fire to significantly higher than the pre-fire levels (Fig. 8.12B). Presumably nitrification was very efficient just after the fire with little NH_4^+ released, while in more recent years nitrification was reduced and the young growing forest is not taking up all the NH_4^+ from mineralization or precipitation.

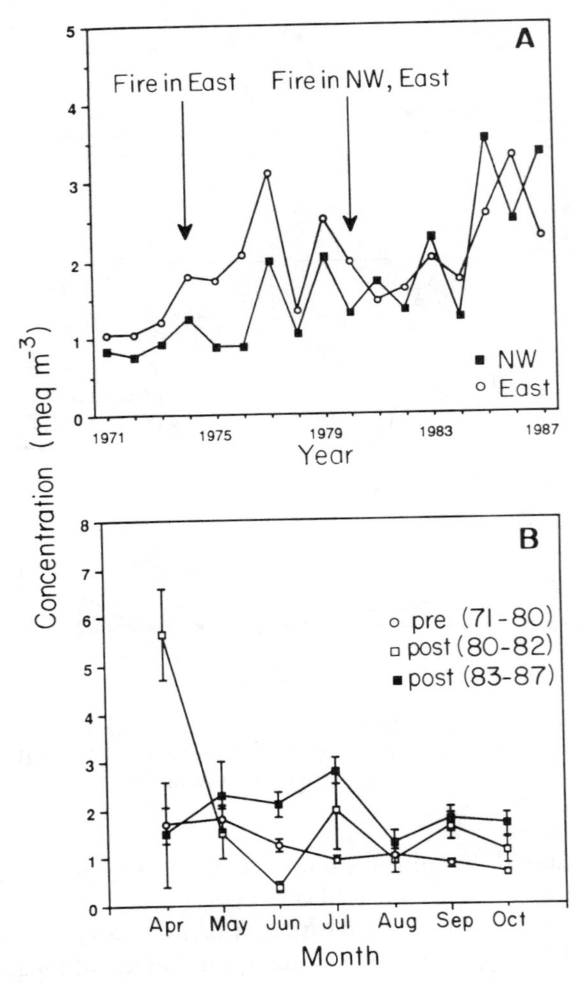

Figure 8.12 Ammonium–nitrogen concentrations in ELA streams. A: Mean annual concentrations in the Northwest and East streams from 1971 to 1987; B: Mean monthly NH_4^+-N concentration in Northwest stream before and after the 1980 fire.

8.5.6 EFFECT OF FIRE ON pH AND ALKALINITY IN STREAMS

Alkalinity and pH in poorly buffered streams are maintained by a complex set of reactions involving base cations, strong acid anions, organic acids, vegetative growth, respiration, nitrification, denitrification and other biological processes. The pH of the NW stream prior to burning averaged 5.45. It decreased after the 1980 fire to average 5.07 in 1981 and 4.99 in 1982 (Fig. 8.13A). Monthly pH values show that the pH decline occurred throughout the

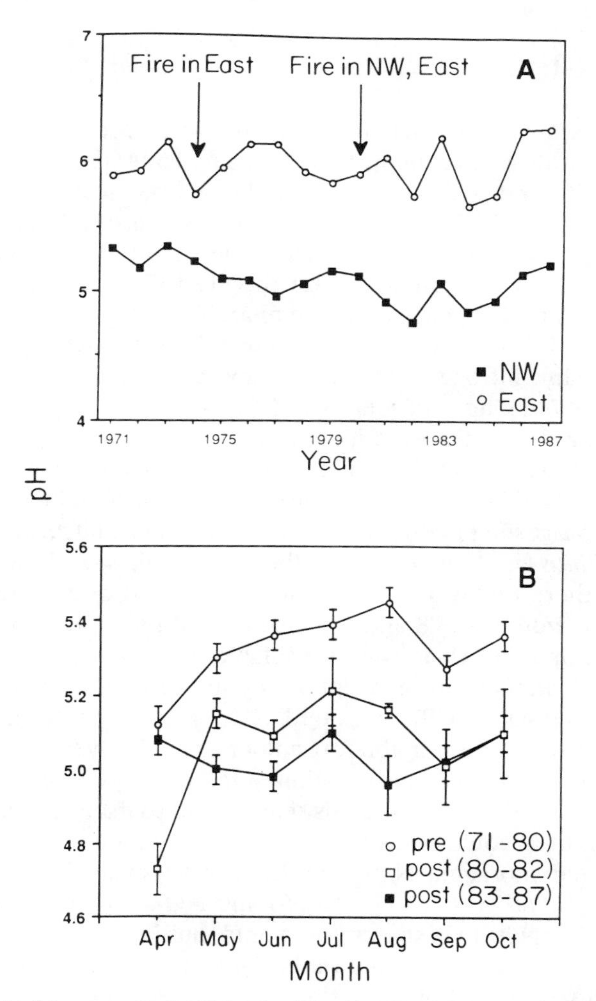

Figure 8.13 A: Mean annual pH in the Northwest and East streams from 1971 to 1987; B: Mean monthly pH in the Northwest stream before and after the 1980 fire.

year, and lasted for several years (Fig. 8.13b). A similar decrease in pH due to fire was detected by Nakane *et al.* (1983) in Japan. This study compared pH in adjacent basins: an unburned forest (6.86) with a moderately burned (6.15) and a severely burned forest (5.74). The alkalinity in their streams decreased in the same fashion (122 vs 24 vs 13 meq m^{-3}). Thus fire (this study and Nakane *et al.*, 1983) and clearcutting (Likens *et al.*, 1970) can both cause a decrease in stream pH. The pH in the E stream did not decrease after fire, perhaps because this basin is more bicarbonate and base cation rich than is the NW Basin.

8.6 DISCUSSION

Ecosystem losses of nutrients in streamflow after fire were low at ELA compared to atmospheric inputs (see later). Total nitrogen losses from volatilization during the ELA fires are unknown but such losses have been reported to be quite high in other conifer ecosystems (900 kg ha^{-1}, Grier, 1975). Losses of total nitrogen by leaching via stream flow in the NW Basin were 1.3 kg ha^{-1} in the first year after the fire (of which 0.55 kg ha^{-1} was dissolved inorganic nitrogen). This is comparable with the 2.1 kg ha^{-1} loss of total N from the E Basin one year after its first fire (of which 0.62 kg ha^{-1} was dissolved inorganic nitrogen). Maximum export after the second fire in the E Basin yielded 1.74 kg ha^{-1} of total N in 1982 two years after the fire year (of which 0.3 kg ha^{-1} was dissolved inorganic N). Thus a second burn (1980) in the E Basin produced less N loss than the first burn but was comparable to the yields from the NW Basin which burned for the first time in 1980.

Foster and Morrison (1976) quantified the amount of nitrogen in a stand of 30 year-old *Pinus banksiana* trees in the Great Lakes–St Lawrence Forest region of northern Ontario. They found 165 kg ha^{-1} of N in the trees, the organic soil horizon was 328 and the mineral soil (gravelly loamy sand) was 3729 kg ha^{-1}, most of which was unavailable for plant uptake. They determined that the annual nitrogen uptake by all vegetation was 32 kg ha^{-1}. Ohmann and Grigal (1979) measured 320 kg ha^{-1} of N in the surface vegetation and 105 kg ha^{-1} in the forest floor of a $\approx$100-year-old forest in the Boundary Waters Canoe area of Minnesota, just south of the ELA site. Mineral soil in the ELA watersheds is shallow and sparsely distributed, so the standing stock of N could be as low as 500 kg ha^{-1} in our mature forests.

Input of total inorganic N to the ELA watersheds from precipitation averaged 4.37 kg ha^{-1} $year^{-1}$ (1971–85) and NO_3^- input averaged 2.25 kg ha^-1 $year^{-1}$. The N lost via streamflow would thus be replaced within the first year. Leaching losses of nutrients were minimal in terms of the N available from the plants, soil, and precipitation.

The newly growing forest and the soil rapidly accumulated nitrogen so that

Table 8.3a Mean growing season concentrations of ions in Northwest stream prior to and after fire (meq l^{-3})

	Pre-fire 1971–80	Post-fire 1980–83	Young forest 1983–87
NO_3-N	0.8	1.9	0.9
NH_3-N	1.00	1.3	1.8
Ca^{2+}	83.3	137.1	118.6
Mg^{2+}	53.4	87.2	66.2
Na^+	44.4	70.1	54.7
K^+	3.1	14.2	8.5
SO_4^{2-}	40.5	194.5	93.4
pH	5.45	5.16	4.95

Table 8.3b Comparison of ELA Northwest watershed stream concentrations with those on Mt Moosilauke, New Hampshire (quoted from Vitousek and Reiners, 1975)

	ELA forest		Mt Moosilauke, New Hampshire	
Elements	Young (3–7 year)	Old (≃100 year)	Logged watersheds (intermediate age)	Unlogged watersheds (old)
NO_3^--N	0.9	0.8	1.8	12
K^+	8.5	3	7	13
Mg^{2+}	66	55	24	40
Ca^{2+}	119	83	36	56
Na^+	55	44	28	29
Cl^-	12	19	13	15

Mean growing season (1 June–30 September) stream concentrations in meq m^{-3}. ELA northwest basin 1971–80 (pre-fire ≃100-year-old forest). Young forest 1983–87 (3–7-year-old forest after fire). Mt Moosilauke, New Hampshire, stream concentrations from five intermediate-aged successional logged versus nine old-aged unlogged forests.

a second burn (six years after the first burn) could still cause a 350% increase in NO_3^- concentration (the first burn caused an 800% increase in NO_3^- concentration). Nitrate concentrations returned to background levels faster than other ions, as might be expected because nitrogen is typically limited in boreal forest ecosystems (Foster and Morrison, 1976). The recovery time for most other ions was long. After four to eight years, concentrations of Ca^{2+}, Mg^{2+}, Na^+, K^+ and SO_4^{2-} were still elevated in the NW stream compared to the pre-fire conditions (Table 8.3a). Vitousek and Reiners (1975) hypothesized that nutrients (and other essential elements) are incorporated rapidly into plant biomass during succession, with nutrient retention reaching a maximum in the 'intermediate'-aged forests and slowly declining to zero retention as the forest matures and a steady state is reached. Their data (Table 8.3b) compare stream nutrient concentrations from an intermediate-aged hardwood forest (previously logged) with an unlogged 'old' forest. Comparison of an ELA stream concentrations from a mature forest ($\approx$100 years) with Vitousek's mature forest stream concentrations show that NO_3^- and K^+ are much lower at ELA than in New Hampshire. Concentrations of Ca^{2+}, Mg^{2+}, Na^+, and Cl^- are slightly higher at ELA than in New Hampshire streams. Export from the very young forests (three to seven years) at ELA showed lower NO_3^- but higher base cation concentrations in stream water than do the intermediate-aged New Hampshire forests. Thus Vitousek and Reiners' hypothesis that very young and very old forest ecosystems will have higher nutrient losses than the intermediate-aged forest is not supported by our ELA forest, where neither the young nor the old forest at ELA loses nitrate. The Vitousek–Reiners hypothesis is partially correct with regard to K loss. The ELA very young forest has higher K loss than the mature forest but losses in the mature forest at ELA are lower than the old New Hampshire forest (streams from the old ELA forest averaged 3 meq m^{-3}, while the New Hampshire streams averaged 13 meq m^{-3}). With regard to base cation losses at ELA, the Vitousek–Reiners hypothesis may be correct. Stream data from both young and old forests show higher base cations losses than the intermediate-aged New Hampshire forest (Table 8.3b). Unfortunately, we do not have data for an intermediate-aged boreal forest, and so cannot directly address their hypotheses.

8.7 SUMMARY

In summary, analysis of fire in five conifer forests shows a great variation in response with respect to loss of elements in stream flow. In all five studies only K^+ concentrations were consistently higher in streams after fire. In the ELA and Little Sioux fires other base cations also increased, but not in the Japanese fire where no changes in Mg^{2+} and Na^+ were detected and Ca^{2+} concentrations decreased after fire. In the Entiat fire, Ca^{2+} concentrations decreased after fire, while there was a slight temporary increase in Mg^{2+} and Na^+. SO_4^{2-} and Cl^- concentrations increased more dramatically than NO_3^- or base cations at ELA, but SO_4^{2-} and Cl^- were not reported in any other studies of wildfire. NO_3^- concentrations increased in three of the studies (ELA, Entiat, and Haney), but no increase was observed in the Little Sioux study. No data were presented in the Japanese fire. We conclude that while soil studies of wildfire may show consistent results with increases in pH, base cations, and NO_3^-, stream studies show much more variable results. Some of the variability may be due to ecological considerations such as the richness of the forest and soils, topography, season of burn, and precipitation after the fire. Some variation in results may be due to differences in type of fire (wildfire vs prescribed fire) and initiation of the time of sampling. Our understanding of the role of fire in managing biogeochemical cycles of nutrients in boreal forests is not advanced. Our understanding of the effects of fire on leaching losses of nitrogen is fairly advanced. There are great gaps in our knowledge about the effects of fire on leaching and erosion of other elements from watersheds, atmospheric losses of elements, release and adsorption of ions in ash and soils, and changes in chemical weathering.

ACKNOWLEDGEMENTS

We gratefully acknowledge the assistance of Brian Parker and Michael Turner with data analysis, Michael Stainton and Garry Linsey for chemical analysis of streamwater, and Ken Beaty for hydrologic data. This study was funded by the Canadian Department of Fisheries and Oceans. The World Wildlife Fund provided funds to assist in the analysis of data.

REFERENCES

APHA (American Public Health Association). (1967) *Standard Methods for the Examination of Water and Wastewater*, American Public Health Association Inc., Washington, DC.

Beaty, K.G. (1981) Hydrometerological data for the Experimental Lakes Area, northwestern Ontario, 1969–1978. Part II. *Can. Data Rep. Fish. Aquat. Sci.* **285**.

Bosch, J.M. and Hewlett, J.D. (1982) A review of catchment experiments to determine the effect of vegetation changes on water yield and evapotranspiration. *J. Hydrol.* **55**, 3–23.

Brinkley, D. and Richter, D. (1987) Nutrient cycles and H^+ budgets of forest ecosystems. *Adv. Ecol. Res.* **16**, 1–51.

Brown, G.W., Gahler, A.R. and Marston, R.B. (1973) Nutrient losses after clear-cut logging and slash burning in the Oregon Coast Range. *Water Resources Res.* **9**, 1450–3.

Brown, W. (1984) Maybe acid rain isn't the villain. *Fortune* **109**, 170–4.

Feller, M.C. and Kimmins, J.P. (1984) Effects of clearcutting and slash burning on streamwater chemistry and watershed nutrient budgets in southwestern British Columbia. *Water Resources Res.* **20**, 29–40.

Foster, N.W. and Morrison, I.K. (1976) Distribution and cycling of nutrients in a natural *Pinus banksiana* ecosystem. *Ecology* **57**, 110–20.

Grier, C.C. (1975) Wildfire effects on nutrient distribution and leaching in a coniferous ecosystem. *Can. J. For. Res.* **5**, 599–607.

Grier, C.C. and Cole, D.W. (1971) Influence of slash burning on ion transport in a forest soil. *Northw. Sci.* **45**, 100–6.

Hetherington, E.D. (1987) The importance of forests in the hydrological regime. *Can. Aquat. Resour. Can. Bull. Fish. Aquat. Sci.* **215**, 179–211.

Johnson, P.L. and Swank, W.T. (1973) Studies of cation budgets in the southern Appalachians on four experimental watersheds with contrasting vegetation. *Ecology* **54**, 70–80.

Kerekes, J., Howell, G. and Pollock, T. (1984) Problems associated with sulphate determination in colored humic waters in Kejimkujik National Park, Nova Scotia (Canada). *Verh. Internat. Verein. Limnol.* **22**, 1811–17.

Krug, E.C. and Frink, C.R. (1983) Acid rain on acid soil: a new perspective. *Science* **221**, 520–5.

Krug, E.C. and Frink, C.R. (1984) Letters. *Science* **225**, 1427–9.

Kusaka, S., Nakane, K. and Mitsudera, M. (1983) Effect of fire on water and major nutrient budgets in forest ecosystems. I. Water balance. *Jap. J. Ecol.* **33**, 323–32.

Likens, G.E. (1969) Nitrification: importance to nutrient losses from a cutover forested ecosystem. *Science* **163**, 1205–6.

Likens, G.E., Bormann, F.H., Johnson, N.M., Fisher, D.W. and Pierce, R.S. (1970) Effects of forest cutting and herbicide treatment on nutrient budgets in the Hubbard Brook watershed-ecosystem. *Ecol. Monogr.* **40**, 23–47.

Likens, G.E., Bormann, F.H., Pierce, R.S., Eaton, J.S. and Johnson, N.M. (1977) *Biogeochemistry of a Forested Ecosystem*, Springer-Verlag, New York.

Marks, P.L. and Bormann, F.H. (1972) Revegetation following forest cutting: mechanisms for return to steady-state nutrient cycling. *Science* **176**, 914–15.

McColl, J.G. and Grigal, D.F. (1977) Nutrient changes following a forest wildfire in Minnesota: effects in watersheds with differing soils. *Oikos* **28**, 105–12.

Nakane, K., Kusaka, S., Mitsudera, M. and Tsubota, H. (1983) Effect of fire on water and major nutrient budgets in forest ecosystems. II. Nutrient balances, input (precipitation) and output (discharge). *Jap. J. Ecol.* **33**, 333–45.

Nilsson, S.N., Miller, H.G. and Miller, J.D. (1982) Forest growth as possible cause of soil and water acidification: an examination of the concepts. *Oikos* **39**, 40–9.

Ohmann, L.F. and Grigal, D.R. (1979) Early revegetation and nutrient dynamics following the 1971 Little Sioux forest fire in northeastern Minnesota. *For. Sci. Monogr.* 21.

Rosenqvist, I.T. (1978) Alternative sources for acidification of river water in Norway. *Sci. Tot. Environ.* **10**, 39–49.

Schindler, D.W., Newbury, R.W., Beaty, K.G., Prokopowich, J., Ruszczynski, T. and Dalton, J.A. (1980) Effects of a windstorm and forest fire on chemical losses from forested watersheds and on the quality of receiving streams. *Can. J. Fish. Aquat. Sci.* **37**, 328–34.

Smith, D.W. (1970) Concentrations of soil nutrients before and after fire. *Can. J. Soil Sci.* **50**, 17–29.

Stainton, M.P., Capel, M.J. and Armstrong, F.A.J. (1977) *The Chemical Analysis of Freshwater*, 2nd edn, Can. Fish. Mar. Ser. Misc. Spec. Publ. 25.

Swank, W.T. (1988) Stream chemistry responses to disturbance. In: Swank, W.T. and Crossley D.A. Jr (Eds), *Forest Hydrology and Ecology at Coweeta*, Springer-Verlag, New York, pp. 339–57.

Tiedemann, A.R., Helvey, J.D. and Anderson, T.D. (1978) Stream chemistry and watershed nutrient economy following wildfire and fertilization in eastern Washington. *J. Environ. Qual.* **7**, 580–8.

Vitousek, P.M. and Reiners, W.A. (1975) Ecosystem succession and nutrient retention: a hypothesis. *BioScience* **25**, 376–81.

Vitousek, P.M., Gosz, J.R., Grier, C.C., Melillo, J.M., Reiners, W.A. and Todd, R.L. (1979) Nitrate losses from disturbed ecosystems. *Science* **204**, 469–74.

Woodmansee, R.G. and Wallach, L.S. (1981) Effects of fire regimes on biogeochemical cycles, pp. 649–69. In: Clark, F.E. and Rosswall, T. (Eds), *Terrestrial Nitrogen Cycles. Ecol. Bull. (Stockholm)*, **33**.

Wright, R.F. (1976) The impact of forest fire on the nutrient influxes to small lakes in northeastern Minnesota. *Ecology* **57**, 649–63.

9 Acidification: Whole-catchment Manipulations

R.F. WRIGHT
Norwegian Institute for Water Research, Box 69, 0808 Oslo, Norway

9.1 INTRODUCTION

Acidification of terrestrial and aquatic ecosystems due to deposition of anthropogenic strong acids from the atmosphere is a widespread regional phenomenon in Europe (Overrein *et al.*, 1980; Wright, 1983) and North America (Schindler, 1988a). Although 'acid rain' research began in the 1800s, intensive large-budget research programs first appeared in the 1970s (Cowling, 1981). Studies of the ecological consequenses of acid deposition first focused on the severity, geographic extent, and history (Almer *et al.*, 1974; Wright and Henriksen, 1978; Wright, 1983; Linthurst *et al.*, 1986). Research then moved towards the identification and examination of the major ecosystem processes that control acidification and cause–effect relationships (Goldstein *et al.*, 1985; Reuss and Johnson, 1986; Reuss *et al.*, 1987). In recent years attention has been given to determining dose–response functions and to predicting future response to changes in acid deposition (Cosby *et al.*, 1985a,b; Gherini *et al.*, 1985). Large-scale experimental manipulations with whole catchments play an increasingly important role in this research.

Controlled acidification experiments have been conducted on lakes, streams, and terrestrial catchments. The best-known whole-lake experiment is the acidification of Lake 223 in the Experimental Lakes Area (ELA) of northwestern Ontario, Canada (Schindler *et al.*, 1985). Streams have been the object of acidification experiments in both North America (Hall *et al.*, 1980; Hall and Likens, 1981) and Europe (Norton *et al.*, 1987). I evaluate here the design and utility of acidification experiments conducted on whole terrestrial catchments in Norway (RAIN project), Sweden (Gårdsjön project), and Canada (ELA peatland experiment). Although other whole-catchment experiments such as the clearfelling experiments at the Hubbard Brook Experimental Forest (Likens *et al.*, 1970; Fuller *et al.*, 1987) shed light on the acidification process, the three projects I discuss here were designed specifically to investigate the ecological effects of acid deposition.

Ecosystem Experiments. Edited by H.A. Mooney *et al.*

9.2 RAIN PROJECT

The goal of the RAIN project (reversing acidification in Norway) is to determine experimentally the effect on water and soil chemistry of changing acid deposition to whole catchments. The project tests the premise that reductions in emissions of SO_2 and NO_x will restore acidified waters. The project comprises two parallel large-scale experimental manipulations—artificial acidification at Sogndal and exclusion of acid rain at Risdalsheia (Fig. 9.1). Treatments began in 1984. Wright *et al.* (1988) present the results for the first four years of treatment (1984–87).

The RAIN project design involves radical change of the chemical climate to whole catchments for the duration of treatment (originally three years, now extended to at least six years) and then monitoring of the return to pre-treatment conditions. During the treatment period the pristine catchments at

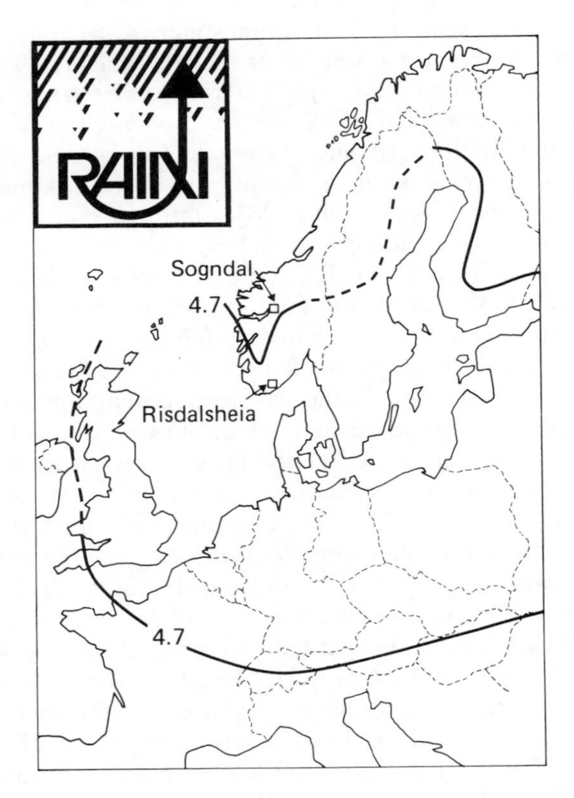

Figure 9.1 Location of the experimental catchments in project RAIN. Areas within the pH 4.7 isoline receive precipitation with a yearly weighted-average pH below 4.7.

Sogndal are acidified while the impacted catchments at Risdalsheia are de-acidified. During the post-treatment period the reverse occurs. The primary project goal is to quantitatively assess the rate of soil and water acidification under controlled conditions and to determine the reversibility and hysteresis of acidification.

9.2.1 SOGNDAL: ACID ADDITION

The Sogndal site is a pristine but sensitive area in western Norway presently receiving only minor acid deposition (precipitation pH 4.8). The site is located 900 m above sea level and characterized by gneissic bedrock, patchy, thin (average depth 30 cm) and poorly developed soils (Lithic Haplumbrept) with pH (H_2O) 4.5–5.5, and alpine vegetation (Lotse and Otabbong, 1985). Four catchments are studied; catchment SOG2 (7220 m^2) receives H_2SO_4, catchment SOG4 (1940 m^2) receives a 1:1 mixture of $H_2SO_4 + HNO_3$, and catchments SOG1 (96 300 m^2) and SOG3 (43 200 m^2) serve as untreated controls (Table 9.1). The H_2SO_4 treatment provides information on the effect of sulfuric acid alone; the $H_2SO_4 + HNO_3$ treatment simulates a future acid deposition scenario in which SO_2 emissions are reduced but NO_x emissions continue unabated. The treated catchments receive identical loadings of acid.

Acid addition began in April 1984 and consists of application to the snowpack of 0.02 mm pH 1.9 acid and four or five events of 11 mm pH 3.2 acid during the snowfree months. Added water is acidified lakewater from SOG1 and is applied at 2 mm/h by means of commercial irrigation equipment. Prior to and following each acid addition 2 mm of unacidified lakewater were added to wet up and wash down the vegetation, respectively (Figure 9.2a). The experiment is designed so that acid addition is obtained with a minimum

Table 9.1 RAIN project: Overview of the experimental catchments and treatments (treatment began April 1984 at Sogndal and June 1984 at Risdalsheia)

Catchment	Treatment	Area (m^2)
Sogndal: acid addition experiments		
SOG1	Control	96 300
SOG2	H_2SO_4	7 220
SOG3	Control	43 200
SOG4	$H_2SO_4 + HNO_3$	1 940
Risdalsheia: acid exclusion experiments		
KIM	Roof, clean rain	860
EGIL	Roof, acid rain	400
ROLF	No roof, acid rain	220

(a)

(b)

Figure 9.2 (a) Application of acid at the RAIN site in Sogndal, Norway. (b) The enclosure at the RAIN site at Risdalsheia, Norway. The roof covers an entire 860-m^2 headwater catchment. Acid precipitation is excluded and clean rain added beneath the roof.

of extra water (about 60 mm added to about 1000 mm natural precipitation) added at an acidity that does not cause direct damage to vegetation. Acid addition was 70 meq/m^3 per year in 1984 and 100 meq/m^2 per year in 1985, 1986, and 1987. Annual acid loading is at levels currently typical for acidified areas in southernmost Norway (Wright *et al.*, 1988).

Results obtained during the first four years of treatment indicate that these pristine, acid-sensitive catchments respond rapidly to increases of acid deposition (Wright *et al.*, 1988). Indeed, a single severe episode of acid precipitation may be sufficient to acidify runoff to the point at which the low pH, high Al water is toxic to fish. These catchments are typical of large regions of southern Norway. Over 50% of Norway is characterized by barren rock and soils less than 70 cm in depth (Overrein *et al.*, 1980). The chronic acidification of runoff at Sogndal during the third and fourth year of treatment is caused entirely by an increase in acid deposition to levels typical for southernmost Norway; neither land-use change or any other environmental change need be invoked.

9.2.2 RISDALSHEIA: ACID EXCLUSION

The Risdalsheia site is located in a sensitive area in southernmost Norway presently receiving high loading of acid deposition (precipitation pH 4.2; sulfate load wet and dry is 100 meq/m^2 per year) (Fig. 9.1). The catchments are situated 300 m above sea level and characterized by exposed granitic bedrock (30–50% of surface) and thin, organic-rich, truncated podzolic soils (average depth 10–15 cm, maximum depth 50 cm) with pH 3.9–4.5, and sparse cover of pine (*Pinus sylvestris* L.) and birch (*Betula pubescens* L.) (Lotse and Otabbong, 1985). Acid exclusion is accomplished by means of a 1200-m^2 transparent roof that completely covers the 860-m^2 KIM catchment (Fig. 9.2b). Incoming precipitation is collected from the roof, pumped through a filter and ion-exchange system, and automatically applied beneath the roof above the canopy by means of a sprinkler system. Seawater salts at ambient levels are re-added prior to sprinkling. Watering rate is fixed at 2 mm/h. An adjacent 400-m^2 catchment (EGIL) has also been covered with a roof and receives ambient acid precipitation by means of an identical sprinkling system. The watering systems are shut down during the winter. Artificial snow is added beneath the roofs each year by means of commercial snow-making equipment. Water from a nearby pond is ion-exchanged and sea salts added to make clean snow at KIM. A third un-covered catchment (ROLF) serves as reference (Table 9.1). Treatment began in June 1984 (Wright *et al.*, 1988).

At Risdalsheia the first 3½ years of acid exclusion show that chemical changes caused by acid deposition are largely reversible (Wright *et al.*, 1988). The mobile strong acid anions SO_4 and NO_3 in runoff respond rapidly to changes in input. At KIM catchment this decrease in mobile anions has been accompanied in part by an increase in alkalinity, but not by a major increase in pH because of buffering by organic acids. The acid-exclusion experiment at Risdalsheia shows that reduction in acid deposition results in decreased concentrations of strong acid anions in runoff. In clearwater lakes and streams in southern Norway a major reduction in the flux of strong acid anions will be reflected in major change in pH and inorganic aluminum; the water will be less toxic to fish and other aquatic organisms.

9.2.3 RAIN PROJECT: DISCUSSION

The changes in runoff chemistry observed at the RAIN project sites can be explained qualitatively by the interaction of key soil–chemical processes such as sulfate adsorption, cation exchange, dissolution of CO_2, mobilization of aluminum, and buffering by organic acids. These processes are central in understanding soil and water acidification by acid deposition (Reuss and Johnson, 1986; Reuss *et al.*, 1987; Brakke *et al.*, 1987). A process-oriented model of soil and water acidification (MAGIC) applied to the first 2-year data generally predicted the observed changes (Wright and Cosby, 1987). The RAIN project provides a unique data set for the calibration and evaluation of such predictive models.

At both Sogndal and Risdalsheia runoff from the treated catchments now differs substantially in chemical composition from that of the control catchments. Runoff at Sogndal has been acidified to levels toxic to fish, and runoff at Risdalsheia has begun to recover to 'pre-acidification' chemical composition. Acid addition at Sogndal has caused soil acidification, whereas acid exclusion at Risdalsheia has initiated reversal of ongoing soil acidification.

The RAIN project is a large-scale demonstration of the role of acid deposition in the acidification of freshwaters (Sogndal) and of the fact that reduction in acid deposition will restore acidified freshwaters (Risdalsheia). It soon became apparent that the exclosures at Risdalsheia were of enormous interest to politicians, environment managers, scientists, environmental interest groups, school groups of all ages, and the general public. Each year about 1000 visitors have found their way to Risdalsheia, despite the fact that it is far from the nearest major city, difficult to find, and accessible only via a private road with locked gate. The visual impact of the roofed catchments has also attracted film and TV crews from several countries in Europe and North America.

9.3 GÅRDSJÖN PROJECT

The Gårdsjön project started in 1979 as an ecosystem approach to study acid deposition and its effects on terrestrial subcatchments and lakes within the Lake Gårdsjön basin (Fig. 9.3). The study is an attempt to quantify the

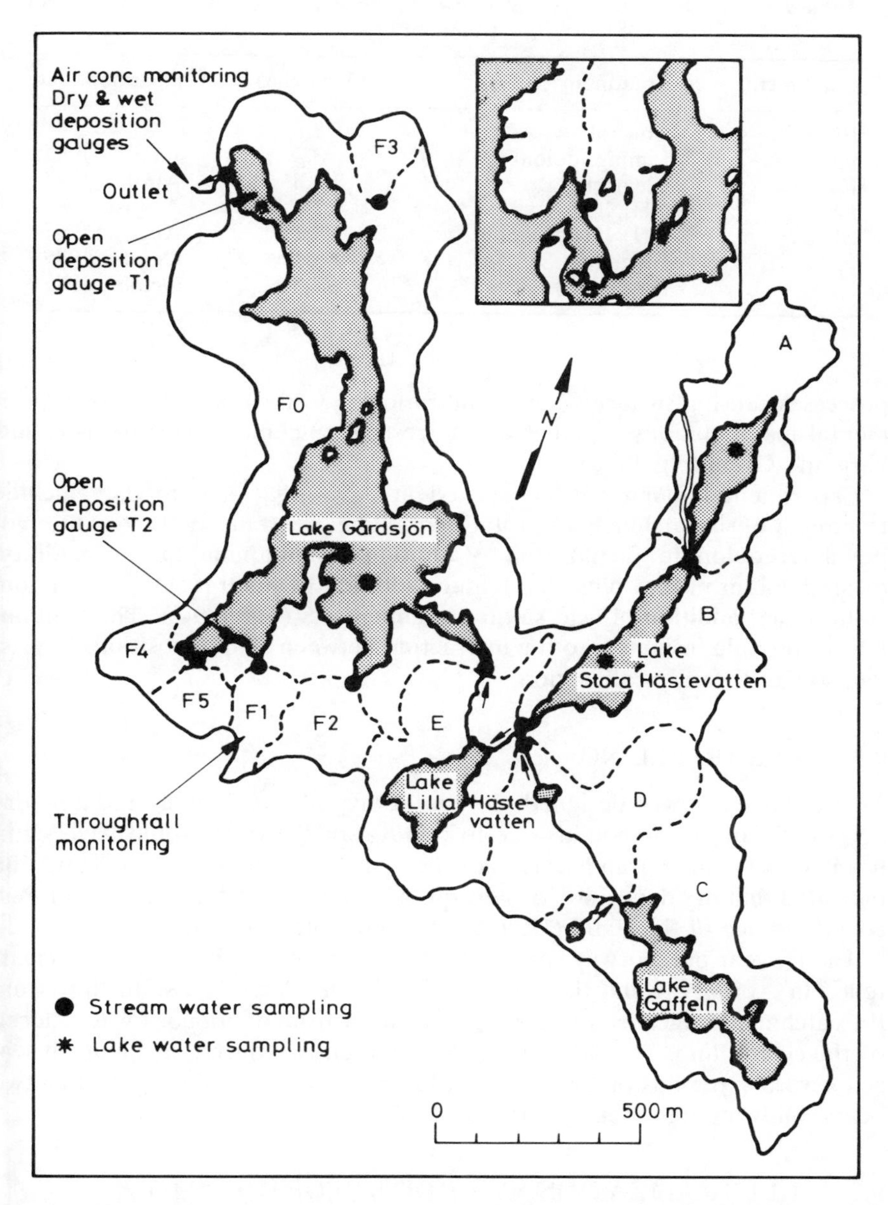

Figure 9.3 Map showing the Lake Gårdsjön catchment on the Swedish west coast and the small subcatchments used for various treatments (from Hultberg and Grennfelt, 1986).

Table 9.2 Gårdsjön project: overview of the experimental catchments and treatments

Catchment	Treatment	Area (ha)	Treatment date
F1	Control	3.6	—
F2	Liming, dolomite	3.3	June 1984
F3	Clearfelling	2.8	April 1984
F4	NH_4NO_3	2.6	August 1984
BE	NH_4NO_3 + liming	4.1	August 1985
F5	Elemental S	3.1	October 1985
L1	Na_2SO_4	2.5	October 1985

processes causing surface water acidification as well as the effects of experimental manipulations on lakes and terrestrial catchment ecosystems (Hultberg and Grennfelt, 1986).

Experimental manipulations started in April 1984 and to date include treatment of six catchments by total clearfelling, neutralization by addition of powdered dolomite, fertilization by addition of ammonium nitrate, addition of ammonium nitrate plus clearfelling, addition of neutral sulfur as sodium sulfate, and addition of acid sulfur as elemental S (Table 9.2). The manipulations provide insights into the interaction between acid deposition, forests, and soil and water acidification.

9.3.1 CLEARFELLING

This experiment was designed to test the hypothesis that the reduced dry deposit of sulfur compounds due to removal of the trees would have significant impact on soil and water acidification. Previous studies at Gårdsjön indicated that dry deposition of sulfur was directly related to amount of forest cover and age of the stand (Hultberg and Greenfelt, 1986).

The 80-year-old Norway spruce stand on catchment F3 (2.8 ha) was clearfelled in April 1984 and the boles removed. The decreased sulfur flux from the catchment indicates that the input of sulfur from dry deposit was reduced by the clearfelling. The clearfelling also apparently reduced the inputs of seasalt aerosols; the flux of Na, Mg and Cl in runoff declined during the first two years following treatment (Hultberg, 1987a).

9.3.2 NEUTRALIZATION BY ADDITION OF DOLOMITE

Inasmuch as water acidification is contingent upon acid soils in the catchment, reduction of soil acidity by addition of neutralizing substances such as carbonate should reduce runoff acidity. This hypothesis was tested in June

1984 when 1.5 tons of finely ground dolomite was spread at catchment F2 (Hultberg and Grennfelt, 1986; Hultberg, 1987a).

9.3.3 FOREST FERTILIZATION BY ADDITION OF AMMONIUM NITRATE

Nitrogen is generally believed to be the growth-limiting nutrient for temporate boreal forests. Fertilization with nitrogen has long been practiced in Sweden to increase forest growth. During the past few years such fertilization has had little or no effect on Norway spruce stands in southern Sweden. Here nitrogen deposition from the atmosphere is substantial, and it is hypothesized that a different nutrient now limits growth. To test this hypothesis catchment F4 was fertilized in August 1984 with ammonium nitrate. A second catchment was treated with ammonium nitrate plus limestone. These treatments are part of a long-term study in which new fertilizing techniques will be tested, and the effects on tree growth and water quality studied (Hultberg and Grennfelt, 1986).

9.3.4 ADDITION OF NEUTRAL AND ACIDIC SULFUR

Sulfur is the dominant component of acid deposition, and sulfate is generally the dominant anion involved in soil and water acidification. The role of sulfur deposition at Gårdsjön has been studied by means of two whole-catchment manipulations. Catchment L1 received 90 kg S/ha and 108 kg S/ha as Na_2SO_4 in October 1985 and October 1986, respectively, while catchment F5 received 112 kg S/ha as elemental S each year (Hultberg, 1987b). This dose is approximately triple the ambient levels of acid deposition at Gårdsjön.

9.3.5 GÅRDSJÖN PROJECT: DISCUSSION

It is the wide variety of different treatments on adjacent catchments that provides the strength of the Gårdsjön experiments. Many of the manipulations have been carried out singly at other sites such as the clearfellings at Hubbard Brook Experimental Forest in 1965 (Likens *et al.*, 1970) and again in 1983–84 (Fuller *et al.*, 1987). Each of the treatments represents a dramatic perturbation of the terrestrial ecosystem, and the observed changes in soil and water chemistry provide information regarding dominant processes and rates. The treatments can thus be interpreted in combination; for example, the dolomite addition and the elemental S addition pushed soil acidity in opposite directions. Also the various treated catchments can make use of the same control catchments. The experimental philosophy for the terrestrial catchments of the Gårdsjön project is thus similar to that for lakes of the Experimental Lakes Area in Canada (Schindler, 1988b).

9.4 ELA PEATLAND ACIDIFICATION

The effect of acid deposition on peatlands has been generally neglected in international acid rain research (Gorham *et al.*, 1984) despite the fact that wetlands are abundant in many of the impacted areas of Scandinavia and eastern North America. Bayley *et al.* (1986, 1987) have carried out an experimental addition of nitric and sulfuric acids to a peatland catchment in the Experimental Lakes Area, Ontario, Canada, to study the effects of acid deposition on *Sphagnum* mosses and on the chemistry of surface and runoff waters. This experiment nicely complements the ongoing whole-lake acidification experiments in the ELA (Schindler, (1988); Schindler *et al.*, 1985).

The experimental design is similar to that used in the Norwegian experiments at Sogndal (the experiments were designed independently at essentially the same time), and includes periodic addition of lakewater acidified to pH 3.0, pre-wetting and post-washing with unacidified water, and a total dose intended to give annual weighted-average pH of precipitation of 4.0. The remarkable similarity of the experimental designs suggests that the compromise between the two major constraints on such large projects (financial resources and ecologically realistic treatment) severely limit the possible experimental design.

9.5 DISCUSSION

Large-scale controlled whole-ecosystem experiments play an increasingly important role in acid rain research. This research field has matured from the initial phase of description of the effects and their geographic distributions to the later phases of dose–response functions and prediction of future trends. Large-scale experiments are particularly relevant in these later phases, both as research tools and as environmental demonstrations that become central in forming national and international environmental policy. There is nothing better than a well-designed and executed whole-ecosystem experiment to convince sceptical scientists and environmental policy-makers of the role of specific pollutants in causing ecological change. Here the ELA experiments on whole-lake eutrophication provide an outstanding example (Schindler, 1974). Repeated visits by UK politicians and environmental and industrial officials to the RAIN project site in Norway and the Gårdsjön site in Sweden apparently played a key role in the 1987 official British change in policy regarding emissions of sulfur and nitrogen oxides.

Whole-catchment experiments in which streamwater is the primary object of interest require hydrologically well-defined catchments sufficiently large in area to give rise to permanent or semi-permanent streams. In humid, temperate regions this means in practice that the catchments be at least 1 ha in area such as those at Gårdsjön. Hydrologically well-defined catchments suitable for experimental manipulation will often have exposed bedrock or very shallow soils around the catchment perimeter. Headwater catchments of small area will thus have very low average soil depth, or lie in steep terrain; otherwise the area will be poorly drained and will become a peatland. These physical restraints mean that the runoff from extremely small catchments may not be chemically representative of large streams in the area. This proved to be a major difficulty in the design of the RAIN experiments at Risdalsheia.

Limited financial resources usually provide the major constraint faced by scientists planning whole-catchment manipulations. The paired-catchment design with appropriate pre-treatment data on the treated catchment has generally been used. This approach entails several trade-offs, all of which are constrained by the financial resources available. For example:

1. Treatment at one large catchment *or* several small replicates?
2. One treatment on a large catchment *or* several treatments on small catchments?
3. One treatment for many years *or* several treatments for one year?

The choices depend on financial resources as well as the expected duration of funding. Nevertheless the more successful whole-catchment experiments have generally been carried out over many years; the value of long-term ecological data has been amply demonstrated by the work at Hubbard Brook (Likens, 1983).

Cost-efficient research can be realized by locating many different experiments in the same geographical region. This strategy has been used successfully with whole-lake experiments at the Experimental Lakes Area in Ontario, Canada (Schindler, 1988b), and with the whole-catchment experiments at Gårdsjön (Hultberg and Grennfelt, 1986). In both cases economy of scale is achieved because of common logistics, common control catchments, common planning and engineering, and the fact that the research personnel generally make the mistakes only once. A second cost-saving approach is to conduct more than one experiment concurrently at the same catchment. This approach presumes that the various treatments do not interfere with one another. An example might be the simultaneous application of acid and a radioctive tracer to study acidification and the fate of radioactive fallout, respectively.

Controlled-ecosystem experiments could clearly play an important role in new research regarding ecological effects of changes in the global atmosphere and climate. Indeed controlled experiments may offer the only realistic approach to quantifying the effects of future changes that have as yet no modern analogs. Examples of such experiments might include doubling the CO_2 concentration in the air above a tropical and boreal forest using exclosures such as at the RAIN project in Norway, and changing the hydrology at a whole catchment to examine the effect on water quality. Results from such experiments will be invaluable in evaluating and further developing predictive models of ecological change.

REFERENCES

Almer, B., Dickson, W., Ekström, C. and Hörnström, E. (1974) Effects of acidification on Swedish lakes. *Ambio* **3**, 30–6.

Bayley, S.E., Behr, R.S. and Kelly, C.A. (1986) Retention and release of S from a freshwater wetland. *Water Air Soil Pollut.* **31**, 101–14.

Bayley, S.E., Vitt, D.H., Newbury, R.W., Beaty, K.G., Behr, R. and Millar, C. (1987) Experimental acidification of a *Sphagnum*-dominated peatland: first year results. *Can. J. Fish. Aquat. Sci.* **44** (Suppl. 1), 192–204.

Brakke, D.F., Norton, S.S. and Henriksen, A. (1987) The relative importance of acidity sources for humic lakes in Norway. *Nature* **329**, 432–4.

Cosby, B.J., Hornberger, G.M., Galloway, J.N. and Wright, R.F. (1985a) Modelling the effects of acid deposition: assessment of a lumped-parameter model of soil water and streamwater chemistry. *Water Resources Res.* **21**, 51–63.

Cosby, B.J., Wright, R.F., Hornberger, G.M. and Galloway, J.N. (1985b) Modelling the effects of acid deposition: estimation of long-term water quality responses in a small forested catchment. *Water Resources Res.* **21**, 1591–603.

Cowling, E.B. (1981) An historical resume of progress in scientific and public understanding of acid precipitation and its biological consequences. Report FR18/1981, SNSF-project, NISK, 1432 Ås, Norway.

Fuller, R.D., Driscoll, C.T., Lawrence, G.B. and Nodvin, S.C. (1987) Processes regulating sulphate flux after whole-tree harvesting. *Nature* **325**, 707–10.

Gherini, S.A., Mok, L., Hudson, R.J.M., Davis, G.F., Chen, C.W. and Goldstein, R.A. (1985) The ILWAS model: formulation and application. *Water Air Soil Pollut.* **26**, 425–59.

Goldstein, R.A., Chen, C.W. and Gherini, S.A. (1985) Integrated lake–watershed acidification study: summary. *Water Air Soil Pollut.* **26**, 327–37.

Gorham, E., Bayley, S.E. and Schindler, D.W. (1984) Ecological effects of acid deposition upon peatlands: a neglected field of 'acid rain' research. *Can. J. Fish. Aquat. Sci.* **41**, 1256–68.

Hall, R.J. and Likens, G.E. (1981) Chemical flux in an acid-stressed stream. *Nature* **292**, 329–31.

Hall, R.J., Likens, G.E., Fiance, S.B. and Hendrey, G.R. (1980) Experimental acidification of a stream in the Hubbard Brook Experimental Forest, New Hampshire. *Ecology* **61**, 976–89.

Hultberg, H. (1987a) Input/output budgets and experimental treatments of acidified catchments in SW Sweden. In: Moldan, B. and Paes, T. (Eds), *GEOMON International Workshop on Geochemistry and Monitoring in Representative Basins*, Geological Survey, Prague, Czechoslovakia, pp. 241–3.

Hultberg, H. (1987b) Experimental treatments of catchments with elemental sulphur and sodium sulfate. In: *Surface Water Acidification Programme, Interim Review Conference, Bergen, Norway*. Royal Society, London, pp. 69–78.

Hultberg, H. and Grennfelt, P. (1986) Gårdsjön project: lake acidification, chemistry in catchment runoff, lake liming and microcatchment manipulations. *Water Air Soil Pollut.* **30**, 31–46.

Likens, G.E. (1983) A priority for ecological research. *Bull. Ecol. Soc. Am.* **64**, 234–43.

Likens, G.E., Bormann, F.H., Johnson, N.M., Fisher, D.W. and Pierce, R.S. (1970) Effects of forest cutting and herbicide treatment on nutrient budgets in the Hubbard Brook watershed-ecosystem. *Ecol. Monogr.* **40**, 23–47.

Linthurst, R.A., Landers, D.H., Eilers, J.M., Kellar, P.E., Brakke, D.F., Overton, W.S., Crowe, R., Meier, E.P., Kanciruk, P. and Jeffries, D.S. (1986) Regional chemical characteristics of lakes in North America. Part II: Eastern United States. *Water Air Soil Pollut.* **31**, 577–91.

Lotse, E. and Otabbong, E. (1985) *Physiochemical properties of soils at Risdalsheia and Sogndal: RAIN project.* Acid Rain Research Report 8/85, Norwegian Institute of Water Research, Oslo.

Norton, S.A., Henriksen, A., Wathne, B.M. and Veidel, A. (1987) Aluminum dynamics in response to experimental additions of acid to a small Norwegian stream. In *Proceedings, Symp. Acidification and Water Pathways*, Vol. 1, Norwegian National Committee for Hydrology, Box 5090 Majorstua, Oslo.

Overrein, L.N., Seip, H.M. and Tollan, A. (1980) *Acid Precipitation – Effects on Forest and Fish.* Report FR 19/80, SNSF-project, 1432 Ås, Norway.

Reuss, J.O. and Johnson, D.W. (1986) *Acid Deposition and the Acidification of Soils and Waters*, Springer-Verlag, New York.

Reuss, J.O., Cosby, B.J. and Wright, R.F. (1987) Chemical processes governing soil and water acidification. *Nature* **329**, 27–32.

Schindler, D.W. (1974) Eutrophication and recovery in experimental lakes: implications for lake management. *Science* **184**, 897–9.

Schindler, D.W. (1988a) Effects of acid rain on freshwater ecosystems. *Science* **239**, 149–57.

Schindler, D.W. (1988b) Experimental studies of chemical stressors on whole lake ecosystems. *Verh. Internat. Verein. Limnol.* **23**, 11–41.

Schindler, D.W., Mills, K.H., Malley, D.F., Findlay, D.L., Shearer, J.A., Davies, I.J., Turner, M.A., Linsey, G.A. and Cruikshank, D. (1985) Long-term ecosystem stress: the effects of years of acidification on a small lake. *Science* **228**, 1395–401.

Wright, R.F. (1983) Acidification of freshwaters in Europe. *Water Qual. Bull.* **8**, 137–42.

Wright, R.F. and Cosby, B.J. (1987) Use of a process-oriented model to predict acidification at manipulated catchments in Norway. *Atmos. Environ.* **21**, 727–30.

Wright, R.F. and Henriksen, A. (1978) Chemistry of small Norwegian lakes, with special reference to acid precipitation. *Limnol. Oceanog.* **23**, 487–98.

Wright, R.F., Lotse, E. and Semb, A. (1988) Reversibility of acidification shown by whole-catchment experiments. *Nature*, **334**, 670–75.

10 Ecosystem Experiments in Wetlands

K.C. EWEL
Department of Forestry, University of Florida, Gainesville, FL 32611–0303, USA

10.1 INTRODUCTION

Interest in studying wetlands has escalated greatly within the last decade, largely in recognition of the ecological and economic damage already suffered because of wetland destruction. In the United States alone, drainage and conversion to agriculture have destroyed more than half the wetlands in the lower 48 states (Tiner, 1984). Sensitivity to these losses in recent years has resulted from improved understanding of the subtle ways in which wetlands affect not only other natural ecosystems but also human societies.

Many wetlands are connected hydrologically to terrestrial and/or aquatic ecosystems; migrating animals, especially birds, connect non-contiguous ecosystems. Recognition of the importance of wetlands in the United States increased dramatically once declines in waterfowl populations migrating to the southern states were associated with increased conversion of freshwater marshes to agriculture in the prairie pothole region in north-central United States and south-central Canada. In other parts of the world declines in fisheries have been associated with wetland destruction, such as removal of mangroves or submersion of floodplains in reservoirs. Therefore, evaluation of the importance of an individual wetland must often encompass a larger region, such as a distant wetland or a downstream water body. This degree of connectedness increases the difficulty of achieving an understanding of the global impacts of wetlands, such as the effect of increased methane release on the ozone layer (summarized by Dickinson and Cicerone, 1986) or the interactions between evapotranspiration (ET) and climate patterns (e.g. Echternacht, 1982).

Although wetland losses have been largely checked in many developed countries, their ecological and economic impacts are still not fully appreciated elsewhere in the world. A major reason for this lag in perception is the necessity of studying wetlands as whole ecosystems before an understanding can be attained of their role within a landscape and a human economy. Such an approach requires simultaneous analysis of flows of carbon, nutrients, and

Ecosystem Experiments. Edited by H.A. Mooney *et al*.

water as they interact with climate and other forcing functions. The ecosystem approach is particularly important in wetlands because of the close coupling among carbon, nutrient, and hydrologic cycles (e.g. Chapin *et al.*, 1980; Howarth and Teal, 1980). Small changes in hydroperiod, nutrient supply, or fire frequency, for example, can have dramatic impacts on other phenomena in a wetland and on the values of the wetland itself. These may be difficult and expensive to quantify, as are the more indirect effects on downstream ecosystems or more remote areas that may be just as important.

An ecosystem analysis may be conducted for any of at least three reasons. First, it may be intended simply to determine how a particular ecosystem functions. At what rate does it fix carbon and accumulate organic matter? What nutrients limit its productivity? What kinds of animals does it produce and support? This basic approach of documentation was a characteristic of the International Biological Program. Today, an ecosystem analysis is more likely to be guided by hypotheses regarding response to perturbations, including explicit management practices such as wastewater disposal, drainage, or timber harvesting, and inadvertent pollution with acid rain or agricultural runoff. Finally, an ecosystem analysis might address hypotheses about the nature of a specific pathway or set of pathways in order to shed light on how ecosystems in general function. This requires a very different approach, but it can be important in interpreting the patterns that are observed in the first two kinds of studies. This chapter describes three kinds of approaches to ecosystem analysis in wetlands in order to provide a basis for determining future research strategies.

10.2 COMPONENTS OF AN ECOSYSTEM ANALYSIS

An ecosystem analysis includes consideration of four important components: carbon, one or more nutrients, water, and at least one forcing function. A modest investment of time and resources is necessary to document standing stocks of carbon, nutrients, and water. Determining turnover of these standing stocks by measuring rates of inflows and outflows is important, but requires more equipment and long monitoring periods. Interactions among these components must be analyzed; for instance, an ecosystem analysis generally includes determining the effect on carbon fixation and allocation of one or more of the other components.

Constructing even a partial carbon budget often dominates an ecosystem analysis. Documenting above-ground plant and animal biomass, species composition and relative abundance, dead organic matter accumulation, above-ground net primary productivity, and decomposition rates are all standard procedures for any complete ecosystem analysis. Details of population dynamics of one or more species of both plants and animals are often also included. Net photosynthesis, respiration, and secondary productivity rates provide more insight into the way an ecosystem functions, but they are much more difficult to obtain.

A hydrologic budget is particularly important to understanding a wetland because of the pervasive influence of reducing conditions that develop when hydroperiods are long, as well as the importance of water-borne nutrient and carbon fluxes. Surveying the basin of a wetland to determine size and depth is necessary to link stage to volume; groundwater inflows and outflows, as well as evapotranspiration, are often poorly documented in a hydrologic budget, making turnover times difficult to calculate.

Constructing budgets of one or more nutrients (usually nitrogen and phosphorus) that affect carbon fixation and allocation requires extensive sampling in water, soil, and organic matter. The prevalence of reducing conditions in wetland soils makes transformations especially important. Uptake and leaching rates are essential for determining turnover times, but they are particularly difficult to measure.

Annual cycles of solar radiation, precipitation, and temperature must be documented in an ecosystem analysis. Fire, drought, and hurricanes are less easily measured, but they can have important effects on many wetlands. Quantification of these effects may require either long monitoring or use of markers such as tree rings or soil cores.

Clearly, a complete analysis that includes only the most straightforward studies outlined here can be extremely expensive, even for a simple ecosystem. Broad-scale pilot studies that address first-order interactions among these components may be necessary to formulate meaningful hypotheses that will then be used to focus the resources of a more intensive project on the most important questions.

10.3 APPROACHES TO ECOSYSTEM ANALYSIS IN WETLANDS

Three different approaches may be taken to a wetland analysis, based largely on the physical size of the wetland but also on the kind of analysis, as outlined above.

1. Wetlands that occupy well-defined basins (or channels) may be the experimental units.
2. Larger or more diffuse wetlands may be better analyzed by conducting replicated studies on small plots.
3. Small basins, or cells, can be constructed within a wetland, allowing both a range of experimental treatments and replication.

In all three approaches the most difficult decisions in establishing a study's protocol involve the trade-off between replication of treatments and establishing a spectrum of response, both within and among sites.

The following sections outline research strategies and results from several ecosystem analyses that are examples of these three approaches.

10.3.1 EXPERIMENTAL UNIT: BASIN WETLAND

Small ponds (<1 ha to ≈12 ha) vegetated by dense stands of cypress trees (*Taxodium distichum*) are common in the southeastern coastal plain of the United States, particularly in Florida, and were evaluated as sites for wastewater recycling. Four swamps were chosen for intensive analysis: two to receive sewage that had been given secondary treatment; one to receive only groundwater, serving as a partial control to distinguish the impacts of nutrient loading and water loading; and one to serve as an undisturbed reference swamp. Six months after the project began, a fire burned one of the sewage swamps and the groundwater swamp. The project therefore continued with an unburned reference swamp, a burned groundwater control swamp, a burned sewage swamp, and an unburned sewage swamp (Table 10.1).

Most studies, particularly intensive metabolism measurements and determinations of water and nutrient budgets, were concentrated in the reference swamp and the unburned sewage swamp. Additional moderately intensive studies were conducted in an undisturbed cypress strand (a larger, slowly flowing swamp) in south Florida and in a strand near Gainesville that had been receiving barely treated sewage for 40 years. Satellite studies examined specific questions in other local cypress ponds as well as in other kinds of swamps where cypress trees grow, such as river swamps and dwarf cypress savannas. The analysis of the response of cypress ponds to wastewater therefore involved virtually no replication of experimental units. Intensive examination of one treated swamp was augmented by analysis of other sewage swamps, not experimentally manipulated but of longer standing.

Certain aspects of the study required using the entire basin as a treatment unit: e.g. impact of nutrients and pathogens on regional groundwater supplies; changes in use of the swamps by animal populations, especially vertebrates; and relative magnitudes of potential nutrient storages. Imposing a treatment that generated a significant response in at least one swamp established two ends of a spectrum (treatment and reference) that allowed major shifts in metabolism and species composition to be detected. The advantage to this approach is the opportunity to collect information on various components within the same treatment unit simultaneously and to observe cross-system interactions. The boundaries of important response variables were established: e.g. changes in net photosynthesis rates (Brown, 1981), decomposition rates (Dierberg and Ewel, 1984), and invertebrate

Table 10.1 Characteristics of cypress swamps used for wastewater treatment analysis

Swamp	Area (ha)	Fire damage	Treatment
Sewage swamp 1	0.5	Severe	Secondary wastewater
Sewage swamp 1	1.0	Slight	Secondary wastewater
Groundwater swamp	0.7	Severe	Groundwater
Reference swamp	4.2	None	None

populations (Brightman, 1984) with nutrient loading; as well as relative importance of soil vs vegetation in nutrient storage and the role of denitrification (Dierberg and Brezonik, 1984).

Another set of projects was subsequently begun to evaluate the impacts on cypress swamps of forest management practices such as drainage, selective logging, and clearcutting. Several swamps in north and central Florida that had already been subjected to such management practices were selected. One study included 15 small swamps in a 4 km^2 management block. Eleven of these swamps had berms and/or drainage ditches affecting their water fluxes to varying degrees, and four were undisturbed (Marois and Ewel, 1983). In another study, three swamps that had been logged at each of five times (from as recently as a few months to as long as 45 years before) were identified (Terwilliger and Ewel, 1986). Although these studies were not such thorough ecosystem analyses as the wastewater project, they built on the information derived from the earlier intensive study to give a more complete picture of how cypress swamps function. The forest management study therefore contained more replication but less intensive analysis of any single swamp.

Natural variability and less dramatic perturbations in swamps that were less intensively studied produced a spectrum of ecological conditions and, in some cases, replicates within single levels in the spectrum. Nevertheless, replication of swamps was essential for establishing the variability of geological substrata (Spangler, 1984), ET rates (Brown, 1981; Heimburg, 1984; Ewel, 1985), and fire effects (Ewel and Mitsch, 1978). Additional swamps incorporated into both the wastewater and silviculture studies provided both replication and intermediate points in the spectrum of nutrient responses. They were useful in suggesting the limits of nutrient loading (Lemlich and Ewel, 1984), establishing vegetation composition as an index to hydroperiod (Marois and Ewel, 1983), and describing variability of net productivity rates within and among swamps (Ewel and Wickenheiser, 1990).

The strategy used to study Florida cypress ponds was to concentrate human and financial resources on a few swamps. In this top-down approach to examining ecosystems, intensive studies on two swamps identified major trends and elucidated important mechanisms, whereas extensive studies on additional swamps determined the variability of many of these responses.

10.3.2 EXPERIMENTAL UNIT: PLOT

Another approach is to come from the bottom up, with detailed studies of specific mechanisms or interactions on separate plots. These results are eventually pieced together, often with the help of a model. This approach is commonly used when a wetland is too large, diffuse, or open to attempt to construct input/output budgets. The most difficult part of this process is the construction of large-scale budgets for the wetland as a whole that outline major regional constraints. For instance, if ET and groundwater fluxes cannot be separated, estimation of water-mediated nutrient and carbon fluxes is difficult.

Plots were used to analyze the major functions of a salt marsh at Sapelo Island in the southeastern United States. Wiegert *et al.* (1981b) outline the long and difficult process involved in identifying and quantifying first major pathways, and eventually the organisms that control them, in the soil, intertidal water column, and emergent shoots. The apparent similarity among salt marshes masks considerable variability in substrate, groundwater flow, and tidal exchange, limiting the ability to draw on studies from other marshes. Attention was focused throughout much of this project on microbial pathways because of the overriding significance of the detritus food web, although primary productivity and herbivory studies were also eventually incorporated.

The major shortcoming of plot studies is extrapolation to the larger scale. For instance, Teal's (1962) data suggested to many that large quantities of organic matter are exported to neighboring estuaries and offshore ecosystems. Later studies using stable carbon isotopes (Haines, 1977; Haines and Montague, 1979) did not support this hypothesis, and it is now likely that the magnitude of export varies from marsh to marsh, depending on both physiographic region and age (e.g. Teal, 1986). Evaluating the nature and magnitude of this pathway in the carbon budget at the Sapelo Island marsh has been assisted by the use of models that integrate the results of the various projects and identify major missing units. This kind of extrapolation has demonstrated that either the scouring effects of storms or microbial assimilation and respiration of detrital carbon could account for the carbon imbalance for this marsh (Wiegert *et al.*, 1981a).

Studies of the tundra at Point Barrow, Alaska (summarized in Brown *et al.*, 1980), also focused on intensive measurements of standing stocks and processes in plots. This group used annual budgets rather than models to integrate the units; models were used in analyzing the individual pathways, however. With this approach, an understanding was developed of nutrient transformations, environmental mechanisms controlling plant productivity, and other major interactions between carbon, nutrient, and water cycles.

However, the scope of these intensive studies was limited. They were concentrated on wet meadow, the second most common of eight vegetation types, accounting for 21% of the research area. Comparisons of wet meadow with other sites were made in each study, and total standing crops and rates for the entire research area were weighted by area of each vegetation type. Extrapolation to the entire landscape that represents the tundra is more difficult, however. Because the slight differences in topography and moisture among the different sites belie significant differences in standing crops and rates, the process of change from one type to another must yet be explored in order to evaluate long-term changes, such as those due to human-induced perturbations (Chapin *et al.*, 1980).

Both the salt marsh and tundra studies have provided detailed information about specific pathways in particular kinds of wetlands. Variability in standing stocks and rates is incorporated in these studies, but there is no guarantee that an understanding of how the whole ecosystem will respond to a perturbation is likely to emerge. Clearly, the trade-off between the basin approach and the plot approach is in scope vs depth. Resources spent on a system-wide carbon budget cannot be spent on species-specific differences in photosynthesis or carbon allocation. Money not spent on travel among sites can be used to increase depth of analysis. A model can be extremely useful in this exercise, especially for establishing priorities for additional research. Unfortunately, model validation is difficult, for the same reasons that prohibit analysis on a scale larger than a plot, and acceptance of the predictions of an unvalidated model may be low.

10.3.3 EXPERIMENTAL UNIT: CELL

Construction of a hydrologically distinct cell within a larger basin is an expensive but often attractive alternative that appears to capture the advantages of both previous methods: the ability to apply treatments to ecosystems with defined boundaries without being overwhelmed by the diversity among them, and the ability to replicate treatments. This approach does have potential disadvantages, however. First, controlling hydroperiod or water level within an experimentally isolated area may cause more or less rapid turnover of water within the cell than actually occurs, depending on how the basins are isolated. Seepage through earthen dikes can generate greater turnover in the water column than is characteristic of the natural ecosystem. Completely impervious walls, on the other hand, may decrease turnover and induce unnatural reducing conditions. A second potential problem with cells is edge effect. For instance, the ratio of perimeter to area is ten times greater in a 10 m × 10 m treatment cell than in a 100 m × 100 m wetland. Finally, the demarcation of boundaries of the cells usually prevents landscape interactions from taking place.

An ecosystem analysis based on cells is being conducted at the Delta Waterfowl Research Station, Manitoba, Canada, where 10 cells of 4–6 ha each were constructed by building dikes in a relatively small area of the large Delta Marsh bordering Lake Manitoba. The purpose of the Delta Marsh project is to quantify the relationship that has been observed between wet–dry cycles and plant and animal productivity (summarized by Weller, 1987). Substrate and seed banks in the marsh remain intact, along with the spatial variability that includes shallow areas and semi-permanent pools. Different treatments can be applied to different cells, however; water stage in each is carefully controlled to provide a replicated spectrum of responses to alternation of drawdown and different levels of flooding (Table 10.2).

Pumping water into small cells to maintain the proper water level seems, in this case, to engender faster turnover than would occur in the larger marsh if it were at the same water level. Although seepage through the dike walls does in fact occur, the first post-treatment nutrient budgets in the Delta Marsh project are dominated by atmospheric nutrient sources (Kadlec, 1986), decreasing the potential importance of different turnover rates in this study.

Edge effects have been encountered in this project. For instance, dead vegetation that accumulates on the surface of the water after extended flooding concentrates in windrows along the northern and western boundaries of the cells (Van der Valk, 1986). This phenomenon also occurs in the undistubed marsh, where the windrows may accumulate in different kinds of areas. Edge effects are largely avoided by sampling at least 10 m from the dikes (A. Van der Valk, personal communication).

Cell construction eliminates serious consideration of landborne transfers of nutrients and carbon across ecosystem boundaries. However, these interac-

Table 10.2 Schedule of water levels for experimental marshes in the Marsh Ecology Research Program at Delta Marsh (after Kadlec, 1986)

Treatment	Number of cells	Year							
		1	2	3	4	5	6	7	8
1	2	N	F	F	D	D	L	L	L
1	3	N	F	F	D	D	M	M	M
1	3	N	F	F	D	D	H	H	H
2	2	N	N	F	F	F	L	L	L
3	2	N	N	N	N	N	N	N	N

N = natural level of marsh (247.5 ± 0.20 m above sea level); F = flooded to 'conditioning' level (248.4 m); D = dry; L = flooded to low level (247.5 m); M = flooded to moderate level (247.8 m); H = flooded to high level (248.1 m).

tions do not appear to be as significant as airborne transfers, such as via insect eclosion and local waterfowl movements.

10.4 CHOOSING AN APPROACH FOR STUDIES OF GLOBAL IMPACTS

All three approaches described above can be used in meeting all three purposes for conducting a wetland analysis. Nor need any analysis be based on only one approach. The tradeoffs among the approaches affect the experimental design, i.e. the number of treatments and the degree of replication. Table 10.3 outlines the tradeoffs involved in establishing the major focuses of the study itself. As the focus of the study narrows from large-scale questions to specific details, the geographic scope of the experimental unit may shrink also. At one end of the spectrum the intact ecosystem should be studied in order to understand how it fits within a landscape. At the other end, details of nutrient cycling that may vary from site to site are best analyzed in small plots. There is obviously considerable overlap among the three, however.

The kind of ecosystem itself may play a significant role in determining the best approach. Many wetlands are, like cypress ponds, well demarcated: peat bogs, prairie potholes, and basin mangrove swamps are good examples. Large peatlands with very slow flow rates, such as lakeshore wetlands, may be better suited for widely spaced plots or cells to allow different treatment levels and replication. Flowing water wetlands have multiple inflows and zones that defy a basin approach but could be better addressed with plots or open-ended cells.

The availability of funds and personnel also influences choice of the best approach. The larger the experimental unit, the more monitoring is needed to identify and track the important interactions that define the ecosystem, and the fewer resources are available to explore the interesting but narrower

Table 10.3 Appropriate levels of study for three approaches to ecosystem analysis

Landscape interactions	Ecosystem budgets	Plant population processes	Vertebrate population processes	Details of nutrient cycling
Basin	Basin		Basin	
		Plot	Plot	Plot
		Cell	Cell	Cell

questions. All the advantages conferred by cells may disappear if funds and space are not available to perform the construction needed.

Questions involving the global impacts of changes in wetland functions can be answered with any or all of these approaches. Either large areas or large numbers of a given type of wetland must be involved. Using cells to replicate a spectrum of treatments, and basins to test landscape-wide impacts of the most important treatments that emerge from preliminary analysis may allow the most effective use of global monitoring systems involving remote sensing and detection of gaseous emissions. These whole-system studies can be backed up by plots in which variability of response of key pathways is determined.

ACKNOWLEDGEMENTS

I thank Drs Arnold Van der Valk and Henry Murkin for providing information about the Delta Marsh Ecology Research Program.

REFERENCES

Brightman, R.S. (1984) Benthic macroinvertebrate response to secondarily treated wastewater in north-central Florida cypress domes. In: Ewel, K.C. and Odum, H.T. (Eds), *Cypress Swamps*, University Presses of Florida, Gainesville, FL, pp. 186–96.

Brown, J., Miller, P.C., Tieszen, L.L. and Bunnell, F.L. (Eds) (1980) *An Arctic Ecosystem: the Coastal Tundra at Barrow, Alaska*. Dowden, Hutchinson & Ross, Stroudsburg, PA.

Brown, S. (1981) Á comparison of the structure, primary productivity, and transpiration of cypress ecosystems in Florida. *Ecol. Monogr.* **51**, 403–27.

Chapin, F.S. III, Miller, P.C., Billings, W.D. and Coyne, P.I. (1980) Carbon and nutrient budgets and their control in coastal tundra. In: Brown, J., Miller, P.C., Tieszen, L.L. and Bunnell, F.L. (Eds), *An Arctic Ecosystem: The Coastal Tundra at Barrow, Alaska*. Dowden, Hutchinson & Ross, Stroudsberg, PA, pp. 458–544.

Dickinson, R.E. and Cicerone, R.J. (1986) Future global warming from atmospheric trace gases. *Nature* **319**, 109–15.

Dierberg, F.E. and Brezonik, P.L. (1984) Nitrogen and phosphorus mass balances in a cypress dome receiving wastewater. In: Ewel, K.C. and Odum, H.T. (Eds), *Cypress Swamps*, University Presses of Florida, Gainsville, FL, pp. 112–18.

Dierberg, F.E. and Ewel, K.C. (1984) The effects of wastewater on decomposition and organic matter accumulation in cypress domes. In: Ewel, K.C. and Odum, H.T. (Eds), *Cypress Swamps*, University Presses of Florida, Gainesville, FL, pp. 164–70.

Echternacht, K.L. (1982) Symposium summary: Regional influence of drainage on the hydrologic cycle in Florida. Coord. Counc. Restor. Kissimmee Rivery Valley and Taylor Creek–Nubbin Slough Basin. Dept. Envir. Regn., Tallahassee, FL.

Ewel, K.C. (1985) Effects of harvesting cypress swamps on water quality and quantity. Fla. Water Resour. Res. Center Publ. 87, Gainesville, FL.

Ewel, K.C. and Mitsch, W.J. (1978) The effects of fire on species composition in cypress dome ecosystems. *Fla. Sci.* **41**, 25–31.

Ewel, K.C. and Wickenheiser, L.P. (1988) Effect of swamp size on growth rates of cypress (*Taxodium distichum*) trees. *Am. Midl. Nat.* **120**, 362–370.
Haines, E.B. (1977) The origins of detritus in Georgia salt marsh estuaries. *Oikos* **29**, 254–60.
Haines, E.B. and Montague, C.L. (1979) Food sources of estuarine invertebrates analyzed using $^{13}C/^{12}C$ ratios. *Ecology* **60**, 48–56.
Harris, L.D. and Vickers, C.R. (1984) Some faunal community characteristics of cypress ponds and the changes induced by perturbations. In: Ewel, K.C. and Odum, H.T. (Eds), *Cypress Swamps*, University Presses of Florida, Gainesville, FL, pp. 171–85.
Heimburg, K. (1984) Hydrology of north-central Florida cypress domes. In: Ewel, K.C. and Odum, H.T. (Eds), *Cypress Swamps*, University Presses of Florida, Gainesville, FL, pp. 72–82.
Howarth, R.W. and Teal, J.M. (1980) Energy flow in a salt marsh ecosystem: the role of reduced inorganic sulfur compounds. *Am. Nat.* **116**, 862–72.
Kadlec, J.A. (1986) Input–output nutrient budgets for small diked marshes. *Can. J. Fish. Aquat. Sci.* **43**, 2009–16.
Lemlich, S.K. and Ewel, K.C. (1984) Effects of wastewater disposal on growth rates of cypress trees. *J. Environ. Qual.* **13**, 602–4.
Marois, K.C. and Ewel, K.C. (1983) Natural and management-related variation in cypress domes. *For. Sci.* **29**, 627–40.
Spangler, D.P. (1984) Geologic variability among six cypress domes in north-central Florida. In: Ewel, K.C. and Odum, H.T. (Eds),*Cypress Swamps*, University Presses of Florida, Gainesville, FL, pp. 60–6.
Teal, J.M. (1962) Energy flow in the salt marsh ecosystem of Georgia. *Ecology* **43**, 614–24.
Teal, J.M. (1986) The ecology of regularly flooded salt marshes of New England: a community profile. *US Fish and Wild. Ser. Biol. Report.* **85**(7.4).
Terwilliger, V.J. and Ewel, K.C. (1986) Regeneration and growth after logging Florida pondcypress domes. *For. Sci.* **32**, 493–506.
Tiner, R.W. (1984) Wetlands of the United States: Current status and recent trends. US Dept. Interior, Fish and Wildl. Serv., Natl. Wetlands Inventory. Sup. Doc., US Government Printing Office, Washington, DC.
Van der Valk, A. (1986) The impact of litter and annual plants on recruitment from the seed bank of a lacustrine wetland. *Aquat. Bot.* **24**, 13–26.
Weller, M.W. (1987)*Freshwater Marshes: Ecology and Wildlife Management*, 2nd edn, University of Minnesota Press, Minneapolis, MN.
Wiegert, R.G., Christian, R.R. and Wetzel, R.L. (1981a) A model view of the marsh. In: Pomeroy, L.R. and Wiegert, R.G. (Eds), *The Ecology of a Salt Marsh*, Springer-Verlag, New York, pp. 183–218.
Wiegert, R.G., Pomeroy, L.R. and Wiebe, W.J. (1981b) Ecology of salt marshes: An introduction. In: Pomeroy, L.R. and Wiegert, R.G. (Eds), *The Ecology of a Salt Marsh*, Springer-Verlag, New York, pp. 3–19.

11 The Disturbance of Forested Watersheds

P.M. ATTIWILL
School of Botany, University of Melbourne, Parkville, Victoria, 3052, Australia

11.1 THE DEVELOPMENT OF WATERSHED STUDIES

> Watershed ecosystems . . . with watertight basins, well-defined watershed boundaries, reasonably homogeneous geologic formations, uniform distributions of soil, vegetation and climate, year-round precipitation and streamflow and several clusters of three or more similar-sized watersheds (provide) ideal conditions where entire watersheds (can) be experimentally treated and compared to gain a clearer understanding about the ecology of . . . landscapes (Likens, 1985).

Here in a nutshell are the aims and requirements of ecological studies of watersheds. There is now an abundance of 'paired catchment' studies, the fundamental aims of which have been to define the effect of management practices—most often the effect of cutting a forest—on water yield. Bosch and Hewlett (1982) brought together the results of 94 such studies. Each 10% change in vegetation cover results, on average, in a 40 mm change in annual water yield in coniferous forest, a 25 mm change in annual water yield in hardwood forest, and a 10 mm change in annual water yield in scrubland.

The Hubbard Brook Ecosystem Study (Likens *et al.*, 1977; Bormann and Likens, 1979) focused our attention on the biotic regulation of hydrological and chemical balances in forested ecosystems. It provided an ecological framework of forest development after disturbance within which ecological studies can be structured. This framework is based on the accumulation of *total* biomass (defined to include the mass of both living and dead biological materials) within a forest ecosystem. It starts with *reorganization* where biomass declines immediately after disturbance. Biomass increases to a maximum during the *aggradation* phase, and declines slightly during *transition* to reach a more or less stable *steady state*.

We might therefore summarize the development of watershed studies as a progression from an emphasis on the effects of disturbance of vegetation on water yield to an emphasis on *biotic regulation after disturbance*—the theme of the work at Hubbard Brook and elsewhere (for example, the disturbance

Ecosystem Experiments. Edited by H.A. Mooney *et al.*

by air pollutants including acidic depositions—see comprehensive review by Binkley and Richter, 1987).

It should be clear from this introduction that the topics of losses and gains of water and of elements including H^+ have been extensively and recently reviewed. The theme of this chapter will be biotic regulation of losses and gains. This theme must inevitably encompass studies at a scale smaller than that of the watershed. The chapter will concentrate on a few selected processes of biotic regulation which seem to be critical to a further understanding of the results of catchment studies.

11.2 WATER YIELD FROM A FORESTED CATCHMENT

Bosch and Hewlett (1982) reported that the only study out of the 94 catchment studies they reviewed which did not fit the general theme of an increase in water yield after disturbance was Langford's (1976) study of *Eucalyptus regnans* (mountain ash) in southeastern Australia. It is worth our while to examine this anomalous result further.

Mountain ash is one of the world's tallest hardwoods and grows on deep soils at altitudes between 500 and 1000 m, with rainfall greater than 1200 mm. It grows in even-aged stands, and seedlings do not establish beneath the mature forest. Fire plays a major role in the evolution and ecology of the mountain ash forest. The trees are killed by fires which scorch the crown. Seedfall following a fire is profuse, and seedlings readily establish on the fire-bared soil at the rate of hundreds of thousands per hectare. Mountain ash is intolerant of competition; numbers are reduced to 10 000–20 000 per hectare by age five years, to 200–300 per hectare by age 50 years and to 20–100 per hectare by age 200 years (old-growth forest).

Because of the high rainfall and excellent structure of the soils (high infiltration rate, high water-holding capacity) moutain ash forests have been extensively reserved as water catchments for the city of Melbourne. These reserves contain significant areas of old-growth mountain ash forest aged about 200 years.

Major fires swept through the southeast of Australia in January 1939, and burnt extensive areas of mountain ash forest, including much of that in the water catchment reserves. The regeneration which followed the fire is called '1939 regrowth'. Langford (1976) showed an average annual reduction in streamflow over the entire catchment (that is, both burnt and unburnt areas, or old-growth and regrowth forests) for a 21-year period after the fire of 21% of average streamflow before the fire (Fig. 11.1). Kuczera (1987) subsequently modelled this decline in detail, and showed reductions in streamflow of up to 50%, the reduction increasing with the proportion of the catchment which is covered by stands of 1939 regrowth (Fig. 11.2).

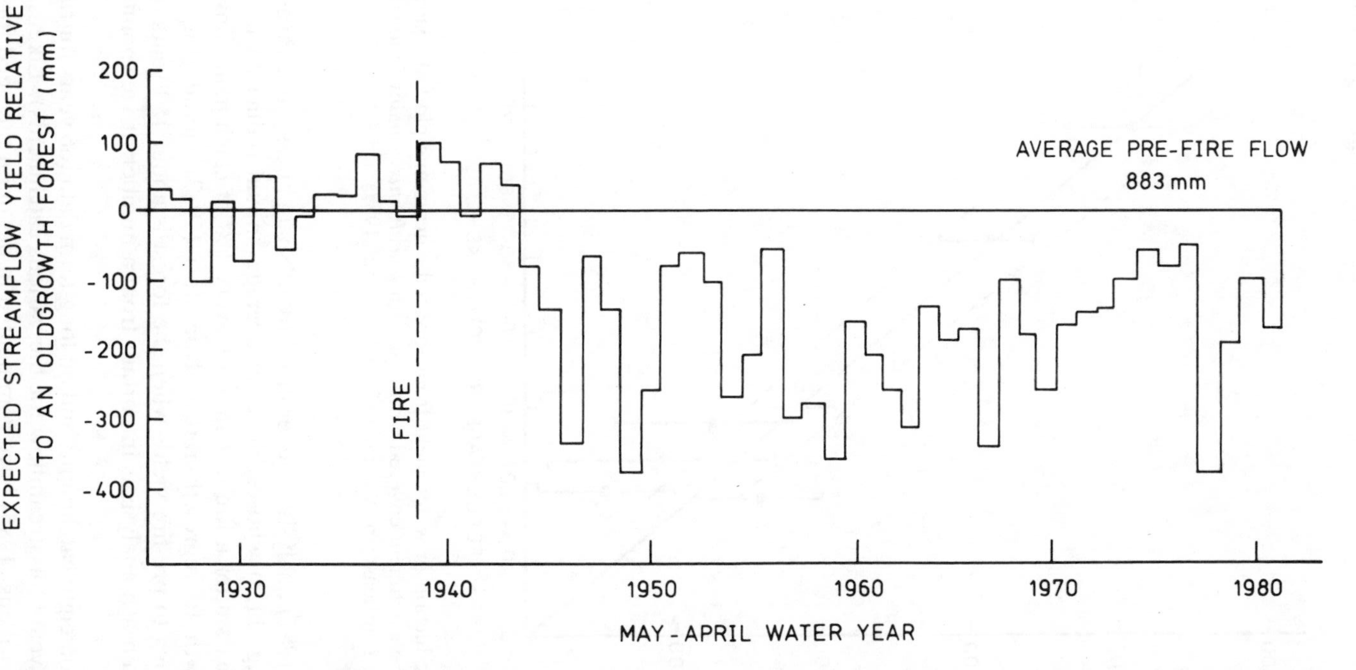

Figure 11.1 Changes in annual streamflow from a forested catchment (*Eucalyptus regnans*) following the 1939 bushfire in southeastern Australia (from Attiwill and Leeper, 1987—after Kuczera, 1987).

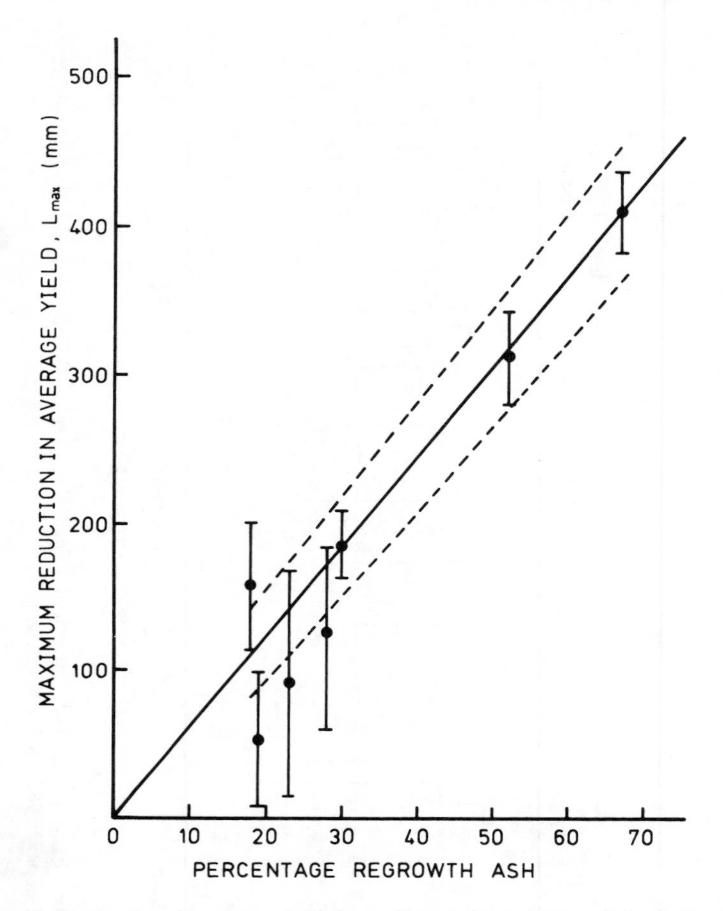

Figure 11.2 Reduction in water yield from forested catchments plotted against the proportion of the catchment covered with regrowth *Eucalyptus regnans* following the 1939 bushfire in southeastern Australia (from Kuczera, 1987).

Kuczera's (1987) analysis is generalized by O'Shaughnessy and Jayasuriya (1987) (see Fig. 11.3) who estimate an average annual reduction in water yield 20–30 years after a fire of 6 mm for every 1% of catchment converted from old-growth to regrowth forest. Kuczera's (1987) model (Fig. 11.3) predicts a return to pre-fire yields when the forest is about 150 years old.

The mountain ash analysis is important from a number of viewpoints:

1. It is based on eight catchments and on long-term data (66 years for rainfall and 50–66 years for streamflow for the various catchments). Likens *et al.* (1984) and Likens (1985) have commented in detail on the need for long-term data if the analysis of comparative studies is to be meaningful.

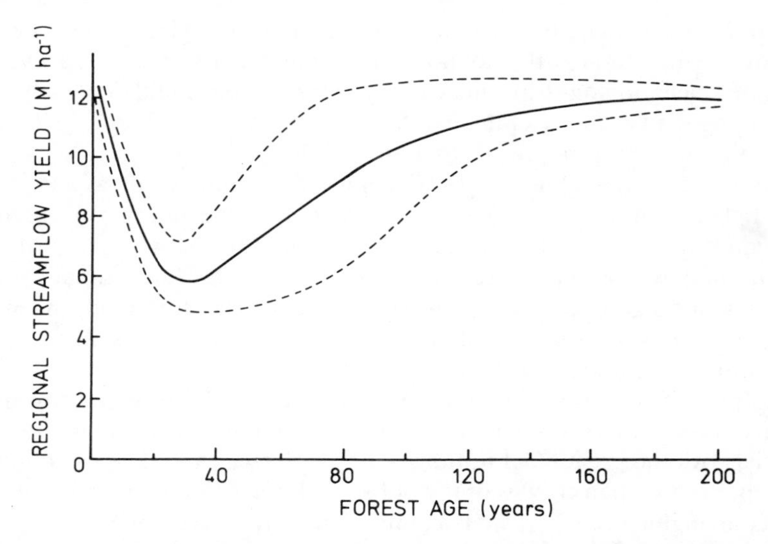

Figure 11.3 Relationship between water yield and stand age for *Eucalyptus regnans* forested catchments (from Kuczera, 1987 and O'Shaughnessy and Jayasuriya, 1987).

2. It is a definitive comparison of old-growth (up to 200 years) and regrowth forest ecosystems (Fig. 11.2)—of an ecosystem which, in the sense of Bormann and Likens (1979), is at steady state with one which is rapidly aggrading. Vitousek and Reiners (1975) have commented on the confusion in watershed studies between the use of 'mature' or 'climax' in a phyto-sociological sense and steady state in the ecosystem sense.
3. It provides a definitive statement to the land management agencies. Clearcutting and regenerating would of course maintain a proportion of the forest in an aggrading phase, thereby decreasing water yield. If the present policy of no logging is maintained the eventual increase in stream-flow from the regrowth ash will be equivalent to 30% of the current annual consumption of water by the city of Melbourne with a population of over 3 million (O'Shaughnessy and Jayasuriya, 1987).

11.2.1 BIOTIC CONTROL OF WATER YIELD

A regrowth forest of mountain ash can be described as rapidly aggrading. We have measured the growth of a regenerating mountain ash forest at Britannia Creek, Victoria, following clearcutting with the logging slash burnt by intense bushfire in the summer of 1983. At age three years the basal area of living trees was 25.5 m^2 ha^{-1} and the stocking of living trees was 11 600 ha^{-1}. The biomass of living trees was 60 tonnes ha^{-1}, with a leaf area index of 5.8. At age five years maximum diameter was 17 cm and maximum height was 17 m.

What is the explanation for a decrease in water yield from regrowth compared with old-growth mountain ash? Langford (1976) cited evidence that differences in fog-drip and canopy interception could not cause such large changes in water yield, and that the explanation therefore lies in a greater rate of transpiration of the regrowth forest.

Moran and O'Shaughnessy (1984) compared a number of aspects of the water balance of nine catchments in which the vegetation was regrowth mountain ash. On a catchment basis the relationship between rate of water loss (streamflow) and basal area of mountain ash trees (an estimator of leaf area) was not significant, most probably because the perennial and woody understorey and shrub layer flourish where tree density is low, thereby contributing to leaf area of the ecosystem.

Langford (1976) discussed some possible explanations of the lower rate of transpiration of the old-growth forest, including the decrease in water potential with the great height of the old-growth trees and the possibility that the deeper more open crowns of the old-growth trees (leaf area index about 2, P.J. O'Shaughnessy, personal communication), together with a well-developed understorey (leaf area index not known) lead to lower transpiration than that from the shallower crowns of the regrowth forest. On the other hand, it might be expected that atmospheric turbulence within the crowns would be greater in the old-growth forest than in the more-dense regrowth forest.

A series of studies including anatomical (Legge, 1985a) and hydraulic (Legge, 1985b; Legge and Connor, 1985) properties of stems, and gradients of water potential with height (Connor *et al.*, 1977) have been published and more are planned. These studies indicate a lower stomatal conductivity in leaves of old-growth compared with regrowth forest which would result in a lower rate of transpiration in old-growth forest. Much more work is required to provide a definitive resolution. The work is logistically difficult—detailed physiological and micrometeorological measurements in tall forests (over 80 m in the old-growth forest) place great demands on funds, equipment, personnel, and analysis and interpretation. It is not unexpected that data for tall forests worldwide are extremely limited.

We should also ask: Does age *per se* control the rate of transpiration? The regrowth forest is rapidly aggrading—biomass increment and rate of nutrient uptake are high during the first 50 years. Individuals in the old-growth steady-state forest are degrading—crowns die and break, and pockets of fungal infection spread through the heartwood. We should then ask the general question: What is the biological limitation to the size of a tree?

The limitation to size in mountain ash might be structural. That is, as a structural limit is reached, there is a break; the tissue is exposed to fungal infection and the vigour of the tree declines. It is also plausible to propose that size becomes limited by a decrease in the ability of water to move through the transpiration pathway. This would cause death of some of the crown, which would then be prone to fungal infection and breakage.

The vessels in the regrowth trees are large and visible in transverse section to the naked eye. When a tree is felled and the leaves are removed, water flows freely from the xylem. Our proposition would be substantiated if the xylem becomes less conductive with age, either by a decrease in size of the vessels or as a result of occlusion of the vessels.

In summary, the definitive study of mountain ash watersheds in Victoria shows a significant reduction in water yield from regrowth forest compared with old-growth forest. There is no proven explanation. The general question —what is the limit to the size of a tree—seems as fundamental to this analysis as it is to others. It is fundamental, for example, to the problem of increasing atmospheric carbon dioxide and the capacity of forests to store more carbon (see Eriksson and Welander, 1956, who first posed the question in this context).

Many of the paired-catchment studies involve the comparison of one catchment which receives no treatment with another which is disturbed (by clearcutting, for example). Likens (1985) made the valuable point that an untreated catchment is not a *control* in the true experimental sense but rather it is a *reference* used for a comparison.

We should analyze more critically in many watershed comparisons whether the chosen reference is valid. The answer depends on the aims of the study. For example, the aim of the mountain ash study on water yield was to determine the difference in water yield between old-growth and young-growth forest, and the old-growth forest is the correct reference. In many of the studies on nutrient budgets following disturbance, however, it seems to be assumed that the undisturbed mature or old-age forest has integrity and stability in perpetuity. The undisturbed forest may be a useful reference primarily because it demonstrates the cycling and conservation of nutrients. If the aim, however, is to assess the effects of management practices such as cutting and burning on nutrient losses associated with regeneration, then the undisturbed forest is not an appropriate reference. Rather, the reference should perhaps be the forest which has been disturbed by the forces of nature (by wildfire, for example).

11.3 THE LEACHING OF IONS FROM FORESTED ECOSYSTEMS

The aim of a number of paired-catchment studies has been to quantify the nutrient losses from a disturbed watershed. The Hubbard Brook Ecosystem Study is a classic example (Fig. 11.4, from Likens, 1985—see also Likens *et al.*, 1977; Bormann and Likens, 1979). Watershed W2 was clearfelled in

1965–66 and kept free of regrowth by herbicide application until the end of 1968. One of the most pronounced effects was the increased flux of $NO_3^- - N$ associated with an increase in nitrification in the deforested ecosystem (Fig. 11.4).

The control of ionic leaching by the supply of mobile anions has been thoroughly reviewed (see in particular Nye and Greenland, 1960; Johnson and Cole, 1980; Vitousek, 1983). The role of the anion NO_3-N is of major interest.

To take figures for a highly productive Australian eucalypt forest (Baker and Attiwill, 1981), for example: the amount of nitrogen in the above-ground stand may be 600 kg ha^{-1}, that in the litter layer 200 kg ha^{-1}, and that in the soil 15 000 kg ha^{-1} (or 95% of total N in the ecosystem). The annual turnover of N in litterfall is 50 kg ha^{-1}, an order of magnitude greater than the input of N in rainwater and one to two orders of magnitude greater than the amount of N lost from the ecosystem in drainage.

The fate of this relatively large turnover of nitrogen from plant to soil is under biological control. The nitrogen is held in our example forest within some 8 t ha^{-1} of annual litterfall. This high ratio of C:N maintains a continuous cycling of N in the microbial biomass (Jansson, 1958; Paul and Juma, 1981) and we have demonstrated the dominance of $NH_4^+ - N$ in all but the most productive and N-rich soils of eucalypt forests we have studied (Adams and Attiwill, 1982, 1986; Adams *et al.*, 1990).

When a forest is clearfelled, rates of decomposition, mineralization, and nitrification increase, and the rate of nutrient uptake becomes temporarily zero. The highly mobile NO_3^- is leached out of the system (Fig. 11.4), taking with it the cations in decreasing order Ca^{2+}, Mg^{2+}, K^+, Na^+ (Weston and Attiwill, submitted).

There has been much interest in the time required for the system to return to equilibrium. The Hubbard Brook study with catchment W2 (Fig. 11.4) deliberately aimed at experimentally delaying recovery by the application of herbicides, and the concentration of $NO_3^- - N$ in streamflow from this catchment remained greater than that in the untreated reference for 6 years (Fig. 11.4). Nevertheless, other studies in New Hampshire (Martin and Pierce, 1980) indicate that normal clearcutting operations cause an increase in the concentration of $NO_3^- - N$ in streamflow for about five years.

We now have a considerable amount of documentation on the production and leaching of $NO_3^- - N$ in disturbed forest. A number of studies (e.g. Vitousek *et al.*, 1982) have used the technique of sampling the soil water by lysimeters. While lysimeters are invaluable, they do not provide an estimate of flux of a given element, but they do enable the quantification of *processes* within small and homogeneous areas. The watershed approach integrates these processes over a much larger, more heterogeneous area (see Vitousek, 1983, for further comments).

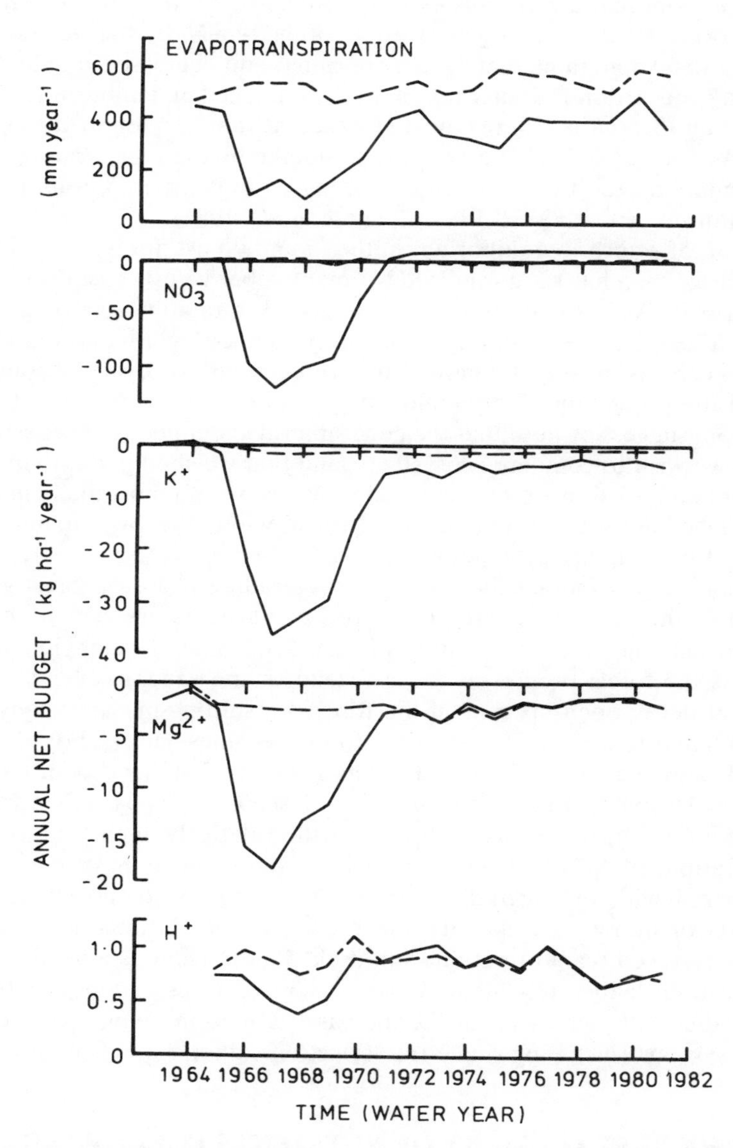

Figure 11.4 Annual net budgets for a forested (dashed line) and experimentally deforested (solid line) catchment at Hubbard Brook Experimental Forest, north-eastern United States (from Likens, 1985).

We have used lysimeters to study the leaching of $NO_3^- - N$ in the clearcut and burnt mountain ash forest described previously in this chapter (Weston and Attiwill, 1990). The concentration of $NO_3^- - N$ in the soil solution increased to 60 ppm immediately after burning, and decreased rapidly thereafter to a concentration similar to that in the unlogged and unburnt reference forest within 18 months. There is no doubt that the rapid depletion of $NO_3^- - N$ in disturbed forest soils is due not only to uptake by the regenerating forest, but to immobilization of nitrogen by soil micro-organisms (Vitousek *et al.*, 1982; Vitousek and Matson, 1984; Polglase *et al.*, 1986).

The rate of return to equilibrium in the mountain ash forest after disturbance is faster than has been reported for many other forests (e.g. Martin and Pierce, 1980). As has been previously discussed, a rapidly aggrading mountain ash forest may result in a decrease in streamflow relative to old-growth forest, in contrast to the Hubbard Brook result where recovery of transpiration and the regulation of streamflow may take decades (Fig. 11.4; Likens, 1985). I am suggesting here that the general term *aggrading* is not specific *per se* when we wish to compare ecosystems, and that we should aim to quantify the *rate* of aggradation. If we were to do this, we would most likely find that the watershed literature is dominated by forests within the lower to mid-range of rates of net primary production.

Processes of nutrient cycling during the aggrading phase of forest growth are *tight* in the sense that large quantities of nutrients are cycled between plant and soil relative to the quantity which is lost from the system in drainage water. As the forest progresses toward steady state, random events such as death and decay open up gaps in the forest—a shifting-mosaic steady state (Bormann and Likens, 1979)—and the forest becomes 'leaky'. For example, Vitousek and Reiners (1975) and Coats *et al.* (1976) have demonstrated relatively low concentrations, fluctuating with season, of $NO_3^- - N$ in streamwater from younger forests compared with relatively high and constant concentrations of $NO_3^- - N$ in streamwater from older forests.

Compared with the aggrading phase after disturbance, however, biotic regulation of nutrient availability and cycling in the shifting-mosaic steady state has received scant attention (Vitousek, 1985). There is a need for such studies within deliberately created gaps in mature forests; opportunities for these studies will increase with the increased attention being given to silvicultural systems (including single-tree selection) other than clearcutting.

11.4 THE REPLACEMENT OF NUTRIENTS FOLLOWING DISTURBANCE

Many of the watershed experiments have given attention to the short-term loss of nutrients from a forest following disturbance by logging. The other

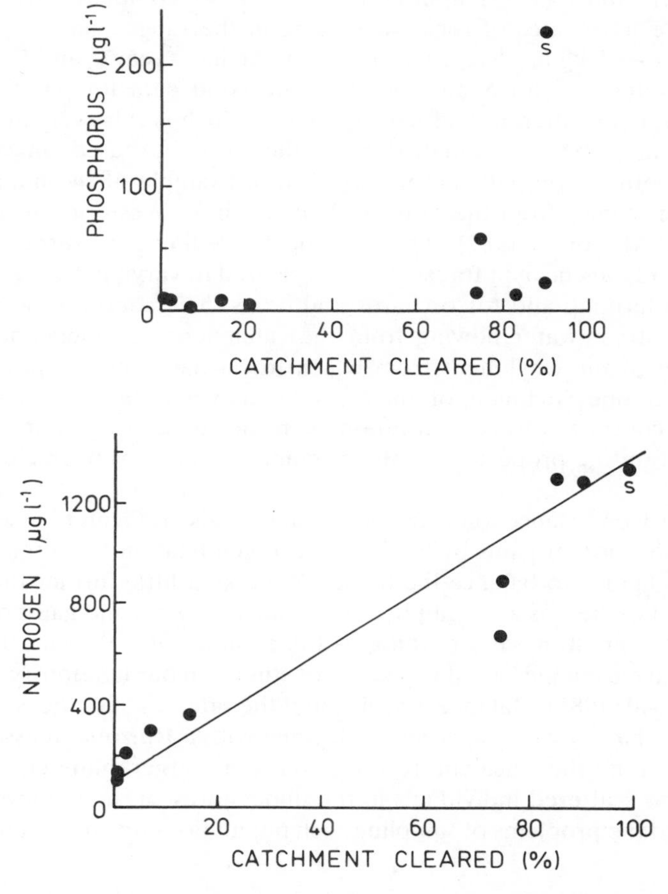

Figure 11.5 Concentration of nitrogen (bottom) and phosphorus (top) in streamwater for subcatchments of the Murray River Basin, Western Australia. The catchments have been cleared to ranging degrees for agriculture and horticulture. All catchments are lateritic except for that marked 'S', which is on a sandy coastal plain (from Attiwill and Leeper, 1987—after McComb, 1982).

side of the balance, of course, is the long-term replenishment from the atmosphere by bulk deposition and biological fixation and from the weathering of rocks.

The most detailed watershed experiments (for example, Likens *et al.*, 1977) have yielded estimates from mass balance studies on the rate of rock weathering. These estimates require precise measurement of nutrient inputs by deposition, of outputs in streamflow, and of changes in the nutrient pools in plants and soil.

From Clayton's (1979) appraisal of the 'most complete' mass balance studies, we have rates of rock weathering in the ranges Na 2–5.8 kg ha^{-1} $year^{-1}$, K 4–7.1 kg ha^{-1} $year^{-1}$, Mg 1.7–52 kg ha^{-1} $year^{-1}$ and Ca trace–86 kg ha^{-1} $year^{-1}$. Unfortunately we have no good data for phosphorus, an element which is often in limiting supply but which is relatively immobile in soil. The mass balance equation must therefore be based on very small numbers both for outputs and inputs. A nice example of the immobility of phosphorus comes from the Murray River Basin in Western Australia (Fig. 11.5, from McComb, 1982). The catchments feeding the Murray River are lateritic, and the eucalypt forest has been cleared to varying degrees for sheep and cattle farming and for orchards and crops. The concentration of phosphorus in streamwater flowing from the catchments is independent of the proportion of the catchment from which forest has been cleared, with the exception of one catchment on the coastal plain where the soil is a sand (Fig. 11.5). In contrast, the concentration of nitrogen in streamwater is highly dependent on the proportion of the catchment which has been cleared (Fig. 11.5).

The nitrogen balance following disturbance is also difficult to assess. Losses by denitrification or gains by biological nitrogen fixation are often presented merely as figures to balance the books. We take a little further the eucalypt forest used earlier, as an example to illustrate the losses and gains of nitrogen following clearcutting and burning with a rotation of 100 years. Inputs and outputs have been measured in a series of studies in our laboratory (Fig. 11.6, from Attiwill, 1985). Balance is achieved if the rate of symbiotic N_2-fixation is about 4 kg ha^{-1} $year^{-1}$. Acacias—a legume with a Rhizobium association—are present in the eucalypt forest initially at larger numbers after fire, reducing to scattered individuals in the understorey of the mature forest. In addition to the problems of sampling such populations on an areal basis, there

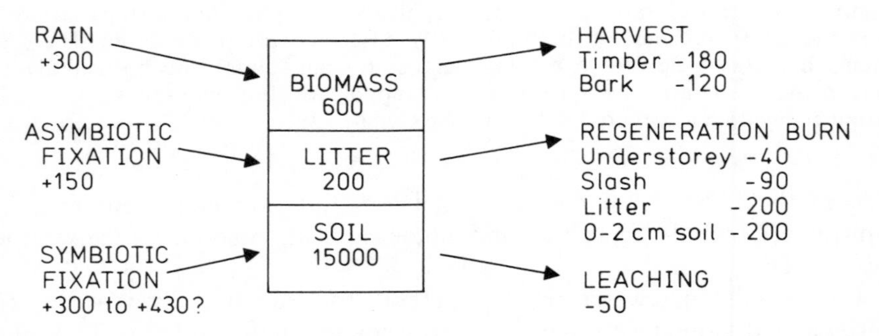

Figure 11.6 Losses and gains of nitrogen (kg ha^{-1} over a 100-year rotation) for a high-rainfall eucalypt forest. The numbers in boxes are pools of N (kg ha^{-1}) (from Attiwill, 1985).

are major methodological difficulties in measuring rates of biological fixation of nitrogen under field conditions (Adams and Attiwill, 1984; Attiwill, 1985). A rate as low as 4 kg ha^{-1} $year^{-1}$ over 100 years to replenish the losses is therefore beyond our resolution.

11.5 BIOLOGICAL REGULATION OF NUTRIENT AVAILABILITY

The availability of nitrogen in the forested ecosystem is biologically controlled. Nitrogen in organic combination in litterfall is mineralized by soil and litter micro-organisms to simple ion, available for uptake (Fig. 11.7).

In situ studies of N-mineralization have now been developed to a useful and generally applicable level (e.g. Rapp *et al.*, 1979; Nadelhoffer *et al.*, 1984, 1985; Adams and Attiwill, 1986; Adams *et al.*, 1989b). These and other studies (e.g. Vitousek and Matson, 1984) have shown that in the undisturbed forest:

1. Rates of mineralization are correlated with measures of productivity and with many independent laboratory-based indices of N-availability.
2. Rates of mineralization are of the same magnitude as the turnover of N in litterfall (Fig. 11.8).
3. Rates of uptake are of the same magnitude as the rates of mineralization; that is, inorganic N does not accumulate in the undisturbed forest. Immobilization of N by soil micro-organisms constantly opposes mineralization so that over many periods of the year net mineralization is negative.
4. The dominant form of inorganic-N available for uptake in many temperate forests is NH_4^+-N.

This knowledge has given us a fundamental understanding of the behaviour of nitrogen when a forest is disturbed. In contrast, our knowledge of phosphorus is much less complete.

The cycle of phosphorus typically recognizes the cycle of P from plant to soil and the biological mineralization of P to simple ion (Fig. 11.9). Yet most of our ecological endeavour has been given to the set of inorganic equilibria—the left-hand side of Figure 11.9.

A major proportion—50% to as much as 80%—of the phosphorus in the surface soil of a forest is organically bound. Only a small proportion of the organic P is, however, available for biological reactions. While we recognize this, we tend in ecological studies to analyse the organic matter in a soil as though it is homogeneous. Of course it is not—if it were, the study of nutrient cycling would be of little relevance since the total amount of nutrients in soil organic matter is large relative to the turnover between plant and soil. Or to

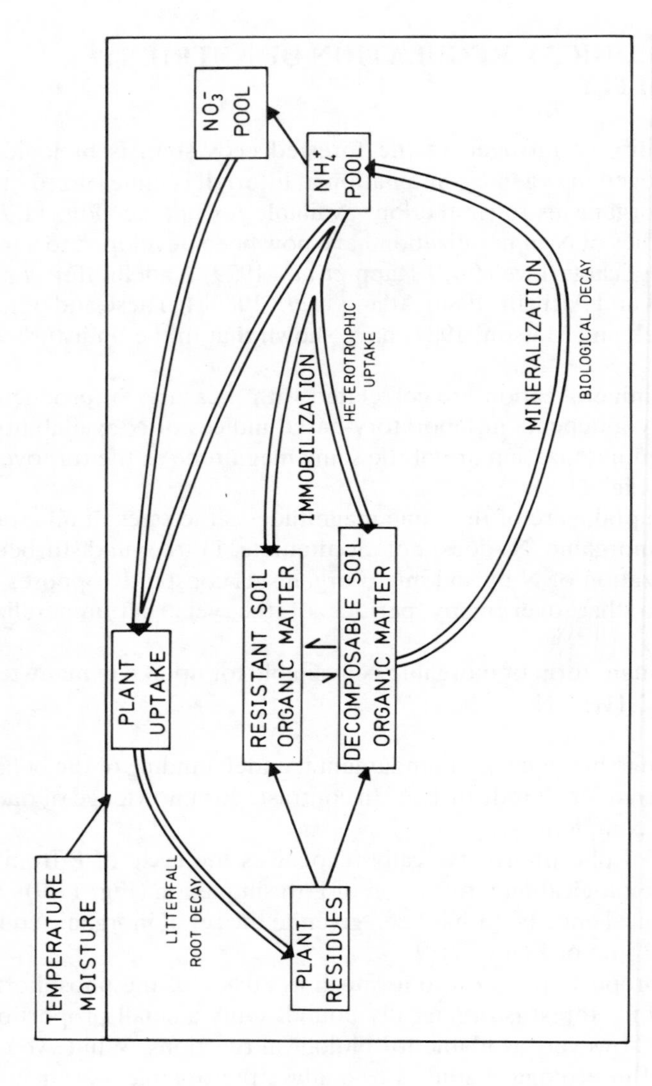

Figure 11.7 A model of nitrogen transformations in a forest soil (from Adams *et al.*, 1989b).

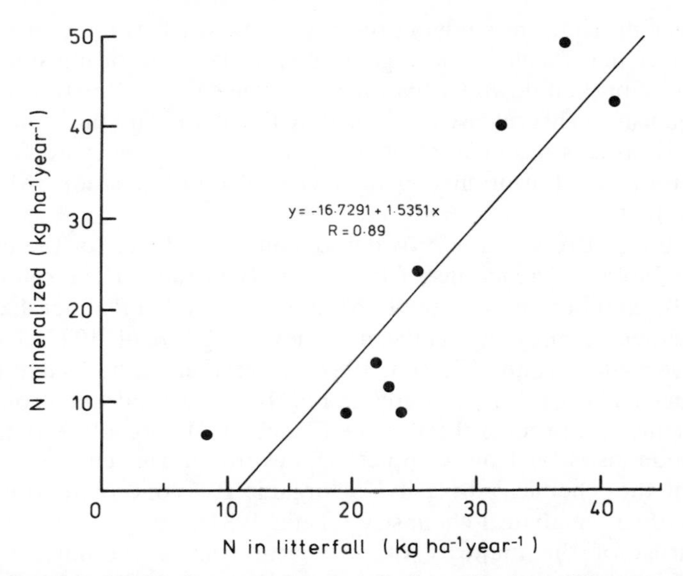

Figure 11.8 Relationship between the rate of mineralization of N measured by *in situ* methods and the rate of return of N in litterfall for a range of eucalypt forests (from Adams *et al.*, 1989b).

turn it around: nutrient cycling is important because the amount of nutrients cycled between plant and soil is large relative to the amount of nutrients held in biologically available soil organic matter. The difficulty is that we cannot separate soil organic matter into discrete homogeneous fractions; empirically, we can separate by extraction (e.g. humic acid, fulvic acid) and conceptually we can separate by chemical reactivity (the old concept of *Nahrhumus* and *Dauerhumus*; see Attiwill and Leeper, 1987).

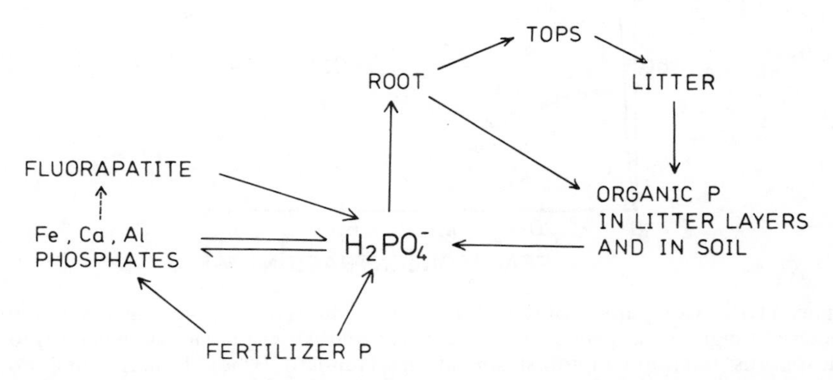

Figure 11.9 The cycle of phosphorus in a forest (from Attiwill and Leeper, 1987).

A further difficulty in studying the rate at which P is mineralized comes from the fact that much of the organic P is in the form of phosphate esters which can be broken down by hydrolase enzymes released extracellularly by micro-organisms. This release is induced by P demand and the mineralization of P by this process has been termed 'biochemical mineralization', to distinguish it from mineralization due to microbial decomposition (McGill and Cole, 1981).

The Hubbard Brook study provided definitive evidence for the overriding control by biological reactions of P availability in the surface soil of forests (Fig. 11.10), and our knowledge of processes involved in P mineralization has been increased recently by significant reviews (e.g. Dalal, 1977; Tate, 1984; Smeck, 1985; Stewart and Tiessen, 1987). Nevertheless, 'rates and pathways of P through soil organic matter are . . . poorly understood when compared to physicochemical aspects of the P cycle' (Tate, 1984). Complete identification of P compounds is far from complete (Stewart and Tiessen, 1987), and the analysis of the 'chemical diversity' of organic P is 'one of the outstanding problems of soil analytical chemistry' (Tate, 1984).

The thrust of the majority of ecological studies continues to be the determination of heterogeneous fractions of organically bound nutrients which, by empirical extraction, show a correlation with rates of forest growth

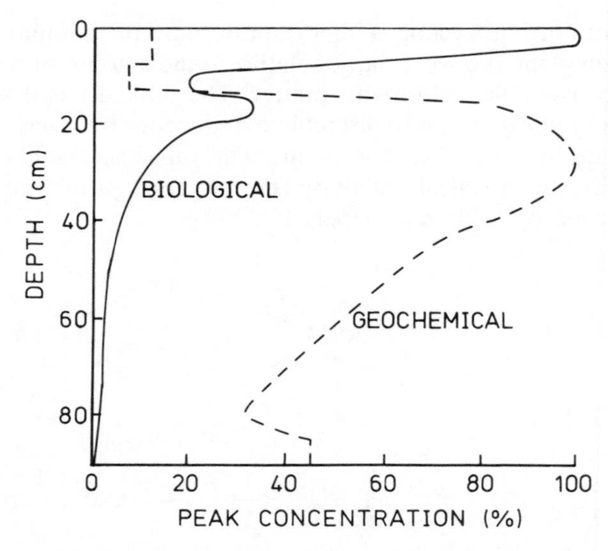

Figure 11.10 Comparison of the relative distributions of biological agents (fine roots, bacteria, fungi) versus geochemical agents (Fe and Al sesquioxides) that contribute to phosphorus retention in forest soil at the Hubbard Brook Experimental Forest, northeastern United States (from Wood *et al.*, 1984).

or nutrient turnover. For example, 'hot salt' extraction of N—boiling the soil in 1 M KCl—has been used extensively, and the N extracted has been termed available N, potentially mineralizable N, potentially available organic N, and total labile N (Oien and Selmer-Olsen, 1980; Whitehead, 1981; Gianello and Bremner, 1986; Adams *et al.*, 1989a). This is indeed a heterogeneous extract which includes inorganic N, N from a number of organic substrates (Gianello and Bremner, 1986), and some microbial-N.

Various extractants for labile organic P have been proposed, the most promising of which perhaps is organic P extracted by 0.5 M $NaHCO_3$ (Olsen *et al.*, 1954; Bowman and Cole, 1978). This fraction includes nucleic acids and derivatives and some phospholipids and other phosphate esters (Bowman and Cole, 1978; Chauhan *et al.*, 1979). Other assays showing most promising results include P held in the microbial biomass and released after fumigation with $CHCl_3$ (Jenkinson and Powlson, 1976; McLaughlin and Alston, 1985; review in Stewart and Tiessen, 1987), and the direct assay of phosphatase activity (Malcolm, 1982).

We have shown, for eucalypt forests over a range of productivity, relatively high correlation between these various indices of organic P availability and between indices of P and totally independent indices of biological availability of N (Adams *et al.*, 1989a). Organic fractions may be much better related to forest productivity than inorganic fractions (Turner and Lambert, 1985; Adams *et al.*, 1989a). Attempts to calculate the rate of P mineralization from changes in labile organic P have been less satisfactory. Variation in the rate is not readily explained in terms of other biological indices (Harrison, 1982); estimates of the net rate in eucalypt forests can be positive or negative, usually close to zero, and are not related to productivity indices such as litterfall (Adams *et al.*, 1989a).

In summary, much remains to be done on the determination and definition of biological control of availability of the two major nutrients—nitrogen and phosphorus—most often listed as limiting the growth of forests. We know rather more about mineralization of N than of P, and reliable *in situ* methods have been developed and proven. Our research in nutrient cycling in forests must increasingly be based on more detailed knowledge of the composition and biochemistry of organic matter in the litter layer and in the soil.

11.6 CONCLUSIONS

1. Long-term data are fundamental to the proper interpretation of watershed studies.
2. We must distinguish between terms used to describe forests in a phytosociological sense and those used to describe the development of ecological processes. In the example for water yield from mountain ash (Fig. 11.3)

steady state is reached at about 200 years and regeneration will result in a decrease in water yield. If, however, the forest aged 100 years was regenerated there would be an increase in water yield. While the 100-year-old trees could be described as 'mature', the ecosystem is clearly not at steady state.

3. In most ecological studies we do not have a *control* in the true experimental sense. Most studies use the untreated forest as a *reference*; in studies of disturbance (for example, by cutting) a better reference may be the forest which has been naturally disturbed (for example, by bushfire).
4. Changes in ecosystem behaviour following disturbance depend on the rate at which the regenerating ecosystem aggrades. This rate should be defined if comparisons between aggrading ecosystems are to be meaningful.
5. Only a small proportion of soil organic matter is biologically reactive and therefore of significance to studies of mineralization and availability. The composition and biochemistry of soil organic matter is fundamental to our understanding of ecosystem behaviour and a much greater emphasis on these aspects is required in ecological studies.
6. Rates of replacement of major nutrients such as phosphorus by chemical weathering and nitrogen by biological N_2 fixation are critical to the long-term balance. Our ability to measure these critical rates is severely limited.
7. Conservation of nitrogen by aggrading ecosystems results from uptake of N by plants and from immobilization of N by soil micro-organisms. A number of workers (e.g. Vitousek *et al.*, 1982) have suggested that the conservation of soil organic matter, thereby maximizing carbon supply to micro-organisms, should therefore be a primary aim in management practices (clearcutting, for example). This argument is sound in principle; it is patently not valid, however, for those forests which, in nature, regenerate only after the major disturbance of intense bushfire.
8. A major proportion of the phosphorus in the surface soil of forests is organically bound. It is most probable that the availability of phosphorus in forests is largely determined by organic equilibria, but our methodology for studying these equilibria is poorly developed.
9. Research in nutrient cycling in forests must increasingly be based on more detailed knowledge of the composition and biochemistry of organic matter in the litter layer and in the soil.

REFERENCES

Adams, M.A. and Attiwill, P.M. (1982) Nitrogen mineralization and nitrate reduction in forests. *Soil Biol. Biochem.* **14**, 197–202.

Adams, M.A. and Attiwill, P.M. (1984) Role of *Acacia* spp. in nutrient balance and cycling in regenerating *Eucalyptus regnans* F. Muell. forests. II. Field studies of acetylene reduction. *Aust. J. Bot.* **32**, 217–23.

Adams, M.A. and Attiwill, P.M. (1986) Nutrient cycling and nitrogen mineralization in eucalypt forests of south-eastern Australia. II. Indices of nitrogen mineralization. *Pl. Soil* **92**, 341–62.

Adams, M.A., Attiwill, P.M. and Polglase, P.J. (1989a) Availability of N and P in forest soils in north-eastern Tasmania. *Biol. Fert. Soils*. **8**, 212–18.

Adams, M.A., Polglase, P.J., Attiwill, P.M. and Weston, C.J. (1989b) *In situ* studies of nitrogen mineralization and uptake in forest soils; some comments on methodology. *Soil Biol. Biochem.* **21**, 423–9.

Attiwill, P.M. (1985) Effects of fire on forest ecosystems. In: Landsberg, J.J. and Parsons, W. (Eds), *Research for Forest Management*, CSIRO, Melbourne, pp. 249–68.

Attiwill, P.M. and Leeper, G.W. (1987) *Forest Soils and Nutrient Cycles*, Melbourne University Press, Melbourne.

Baker, T.G. and Attiwill, P.M. (1981) Nitrogen in Australian eucalypt forests. In: *Australian Forest Nutrition Workshop: Productivity in Perpetuity*, CSIRO, Melbourne, pp. 159–72.

Binkley, D. and Richter, D. (1987) Nutrient cycles and H^+ budgets of forest ecosystems. *Adv. Ecol. Res.* **16**, 1–51.

Bormann, F.H. and Likens, G.E. (1979) *Pattern and Process in a Forested Ecosystem*, Springer-Verlag, New York, Heidelberg and Berlin.

Bosch, J.M. and Hewlett, J.D. (1982) A review of catchment experiments to determine the effect of vegetation changes on water yield and evapotranspiration. *J. Hydrol.* **55**, 3–23.

Bowman, R.A. and Cole, C.V. (1978) Transformations of organic phosphorus substrates in soils as evaluated by $NaHCO_3$ extraction. *Soil Sci.* **125**, 49–54.

Chauhan, B.S., Stewart, J.W.B. and Paul, E.A. (1979) Effect of carbon additions on soil labile inorganic, organic and microbially held phosphate. *Can. J. Soil Sci.* **59**, 387–96.

Clayton, J.L. (1979) Nutrient supply to soil by rock weathering. In: *Proceedings: Impact of Intensive Harvesting on Forest Nutrient Cycling*, State University of New York, Syracuse, New York, pp. 75–96.

Coats, R.M., Leonard, R.L. and Goldman, C.R. (1976) Nitrogen uptake and release in a forested watershed, Lake Tahoe Basin, California. *Ecology* **57**, 995–1004.

Connor, D.J., Legge, N.J. and Turner, N.C. (1977) Water relations of mountain ash (*Eucalyptus regnans* F. Muell.) forests. *Aust. J. Pl. Physiol.* **4**, 753–62.

Dalal, R.C. (1977) Soil organic phosphorus. *Adv. Agron.* **29**, 83–117.

Eriksson, E. and Welander, P. (1956) On a mathematical model of the carbon cycle in nature. *Tellus* **8**, 155–75.

Gianello, C. and Bremner, J.M. (1986) A single chemical method of assessing potentially available organic nitrogen in soil. *Commun. Soil Sci. Pl. Anal.* **17**, 195–214.

Harrison, A.F. (1982) Labile organic phosphorus mineralization in relationship to soil properties. *Soil Biol. Biochem.* **14**, 343–51.

Jansson, S.L. (1958) Tracer studies on nitrogen transformations in soil with special attention to mineralisation-immobilisation relationships. *Kungl. Lantbrükshögsk. Ann.* **24**, 101–61.

Jenkinson, D.S. and Powlson, D.S. (1976) The effects of biocidal treatments on metabolism in soil. V. A method for measuring soil biomass. *Soil Biol. Biochem.* **8**, 209–13.

Johnson, D.W. and Cole, D.W. (1980) Anion mobility in soils: relevance to nutrient transport from forest ecosystems. *Environ. Int.* **3**, 79–90.

Kuczera, G. (1987) Prediction of water yield reductions following a bushfire in ash-mixed species eucalypt forest. *J. Hydrol.* **94**, 215–36.

Langford, K.J. (1976) Change in yield of water following a bushfire in a forest of *Eucalyptus regnans*. *J. Hydrol.* **29**, 87–114.

Legge, N.J. (1985a) Anatomical aspects of water movement through stems of mountain ash (*Eucalyptus regnans* F. Muell.). *Aust. J. Bot.* **33**, 287–98.

Legge, N.J. (1985b) Relating water potential gradients in mountain ash (*Eucalyptus regnans* F. Muell.) to transpiration rate. *Aust. J. Pl. Physiol.* **12**, 89–96.

Legge, N.J. and Connor, D.J. (1985) Hydraulic characteristics of mountain ash *Eucalyptus regnans* (F. Muell.) derived from *in situ* measurements of stem water potential. *Aust. J. Pl. Physiol.* **12**, 77–88.

Likens, G.E. (1985) An experimental approach for the study of ecosystems. *J. Ecol.* **73**, 381–96.

Likens, G.E., Bormann, F.H., Pierce, R.S., Eaton, J.S. and Johnson, N.M. (1977) *Biogeochemistry of a Forested Ecosystem*, Springer-Verlag, New York, Heidelberg and Berlin.

Likens, G.E., Bormann, F.H., Pierce, R.S., Eaton, J.S. and Munn, R.E. (1984) Long-term trends in precipitation chemistry at Hubbard Brook, New Hampshire. *Atmos. Environ.* **18**, 2641–7.

Malcolm, R.E. (1983) Assessment of phosphate activity in soils. *Soil Biol. Biochem.* **15**, 403–8.

Martin, C.W. and Pierce, R.S. (1980) Clearcutting patterns affect nitrate and calcium in streams of New Hampshire. *J. For.* **78**, 268–72.

McComb, A.J. (1982) The effect of land use in catchments in aquatic ecosystems: a case study from Western Australia. *Aust. Soc. Limnol.* Bulletin **9**, 1–24.

McGill, W.B. and Cole, C.V. (1981) Comparative aspects of cycling of organic C, N, S and P through soil organic matter. *Geoderma* **26**, 276–86.

McLaughlin, M.J. and Alston, A.M. (1985) Measurement of phosphorus in the soil microbial biomass: Influence of plant material. *Soil Biol. Biochem.* **17**, 271–4.

Moran, R.J. and O'Shaughnessy, P.J. (1984) Determination of the evapotranspiration of *E. regnans* forested catchments using hydrological measurements. *Agric. Water Manage.* **8**, 57–76.

Nadelhoffer, K.J., Aber, J.D. and Melillo, J.M. (1984) Seasonal patterns of ammonium and nitrate uptake in nine forest ecosystems. *Pl. Soil* **80**, 321–35.

Nadelhoffer, K.J., Aber, J.D. and Melillo, J.M. (1985) Fine roots, net primary production and soil nitrogen availability: A new hypothesis. *Ecology* **66**, 1377–90.

Nye, P.H. and Greenland, D.J. (1960) *The Soil under Shifting Cultivation*, Commonwealth Bureau of Soils Technical Bulletin No. 51.

Oien, A. and Selmer-Olsen, A.R. (1980) A laboratory method for evaluation of available nitrogen in soil. *Acta Agric. Scand.* **30**, 149–56.

Olsen, S.R., Cole, C.V., Watanabe, F.S. and Dean, L.A. (1954) Estimation of available phosphorus in soils by extraction with sodium bicarbonates. United States Department of Agriculture, Circular 939.

O'Shaughnessy, P.J. and Jayasuriya, M.D.A. (1987) Managing the ash type forests for water production in Victoria. *Forest Management in Australia*, Proceedings 1987 Conference Institute of Foresters of Australia, Perth, pp. 437–463.

Paul, E.A. and Juma, N.G. (1981) Mineralization and immobilization of soil nitrogen by microorganisms. In: Clark, F.E. and Rosswall, T. (Eds), *Terrestrial Nitrogen Cycles—Processes, Ecosystem Strategies and Management Impacts*, Ecological Bulletins (Stockholm) **33**, 179–95.

Polglase, P.J., Attiwill, P.M. and Adams, M.A. (1986) Immobilization of soil nitrogen following wildfire in two eucalypt forests of south-eastern Australia. *Acta Oecologica, Oecol. Plant.* **7**, 261–71.

Rapp, M., Le Clerc, M.Cl. and Lossaint, P. (1979) The nitrogen economy in a *Pinus pinea* L. stand. *For. Ecol. Manage.* **2**, 221–31.

Smeck, N.E. (1985) Phosphorus dynamics in soils and landscapes. *Geoderma* **36**, 185–99.

Stewart, J.W.B. and Tiessen, H. (1987) Dynamics of soil organic phosphorus. *Biogeochemistry* **4**, 41–60.

Tate, K.R. (1984) The biological transformation of P in soil. *Pl. Soil* **76**, 245–56.

Turner, J. and Lambert, M.J. (1985) Soil phosphorus forms and related tree growth in a long term *Pinus radiata* phosphate fertilizer trial. *Commun. Soil Sci. Pl. Anal.* **16**, 275–88.

Vitousek, P.M. (1983) Mechanisms of ion leaching in natural and managed ecosystems. In: Mooney, H.A. and Godron, M. (Eds), *Disturbance and Ecosystems—Components of Response*, Springer-Verlag, Berlin, Heidelberg, New York and Tokyo, pp. 129–44.

Vitousek, P.M. (1985) Community turnover and ecosystem nutrient dynamics. In Pickett, S.T.A. and White, P.S. (Eds), *The Ecology of Natural Disturbance and Patch Dynamics*, Academic Press, New York, pp. 325–33.

Vitousek, P.M. and Matson, P.A. (1984) Mechanisms of nitrogen retention in forest ecosystems: a field experiment. *Science* **225**, 51–2.

Vitousek, P.M. and Reiners, W.A. (1975) Ecosystem succession and nutrient retention: a hypothesis. *Biol. Sci.* **25**, 376–81.

Vitousek, P.M., Oosz, J.R., Grier, C.C., Melillo, J.M. and Reiners, W.A. (1982) A comparative analysis of potential nitrification and nitrate mobility in forest ecosystems. *Ecol. Monogr.* **52**, 155–77.

Weston, C.J. and Attiwill, P.M. (1990) Effects of fire and harvesting on nitrogen transformations and ionic mobility in soils of *Eucalyptus regnans* forests of south-eastern Australia. *Oecologia.* **83**, 20–6.

Whitehead, D.C. (1981) An improved chemical extraction method for predicting the supply of available soil nitrogen. *J. Sci. Food Agric.* **32**, 359–65.

Wood, T., Bormann, F.H. and Voigt, G.K. (1984) Phosphorus cycling in a northern hardwood forest: Biological and chemical control. *Science* **223**, 391–3.

12 The Mathematics of Complex Systems

S.A. LEVIN
Section of Ecology and Systematics, Cornell University, Ithaca, New York 14853, USA

12.1 INTRODUCTION

It is not difficult to sell the notion, especially to ecologists, that the world is a complex place, and that the modeling of any ecological system automatically brings one into the realm of complex systems. Indeed, one can advance this idea without ever having to define what is meant by a complex system, a definition that one may feel must be complex itself. Nonetheless, for progress to be made in modeling complex systems, the definitional problem must be confronted.

There is more than one pathway to complexity. Indeed, systems may be complex because their structures are complex, replete with details and interlocking effects; or they may be complex because their dynamics are complex, which may be true even for very simple models. These two notions obviously are not the same; yet either may be implied when one refers to complex systems.

The study of how complicated dynamics can arise from simple models is one of the central areas of research in science today (Gleick, 1988); it is also a vibrant area of mathematical research. Physicists focus on the problem of how turbulence arises from instabilities in fairly regular flows; population biologists and epidemiologists examine similar problems in the dynamics of interacting populations (May, 1974, 1976; Schaffer and Kot, 1985; Schaffer *et al.*, 1985). The stimulus for this revolution came from the work of Edward Lorenz, a meteorologist interested in the (structurally) complex models used to predict the weather. Working on a primitive electronic computer, the Royal McBee, Lorenz in 1960 created a very simplified, three-component model that captured the essential features of the giant weather models. Working with this simple model, Lorenz was surprised to find difficulty in obtaining reproducible predictions of weather patterns. The slightest change in the conditions under which he initiated his simulations would produce dramatically different predictions (Lorenz, 1963, 1964).

Ecosystem Experiments. Edited by H.A. Mooney *et al.*

This property, extreme sensitivity to initial conditions, is a fundamental characteristic of the phenomenon that has been termed chaos, and obviously frustrates our ability to make precise predictions. Chaotic systems typically do not exemplify the fairly regular long-term (asymptotic) dynamics seen in most classical ecological models. The equilibrium points and limit cycles of those models give way to quasiperiodic solutions, and worse yet to structures aptly named *strange attractors*, which wander through space without apparent pattern, making abrupt and unpredictable digressions. Often, these patterns can be understood as the result of competition among distinct and incompatible periods in the system's dynamics, as can occur for example when an inherently oscillatory system is driven (*forced*) by a process (e.g. seasonal demography) with a distinct periodicity. Yet chaotic patterns can arise in completely autonomous systems, that is in systems that are not driven from outside; and it is this feature that has captured the imagination of large groups of theoretical physicists, mathematicians, and other scientists.

Chaotic systems are not without their regularities. Although their time traces may appear indistinguishable from the time-series that would arise from random processes, they exhibit characteristic statistical properties. Modern methods in dynamical systems theory can be used to retrieve the deterministic relations underlying the observed relations (Ruelle and Takens, 1971). Methods of this sort also have been used on ecological and evolutionary data sets in an attempt to extract patterns (Schaffer *et al.*, 1985).

Fractals represent a second active area of research in science and mathematics, with strong relationships to the theory of chaos. Fractals are the creation of Benoit Mandelbrot, a mathematician at IBM, and refer to the complex spatial structures that seem to be everywhere in nature, and that can be generated from very simple theoretical relationships. Fractals turn our attention to the importance of scale, through their elucidation of the fact that the problem of measurement is inextricably tied up with the problem of scale. What has been most seductive here is the observation that, in many applications, there are puzzling regularities in the way measurements vary across scales.

A case in point, perhaps the best-known example, is exhibited by Richardson's data on the lengths of coastlines. As Mandelbrot (1977) discusses, the notion of the length of coastline makes no sense without reference to the precision one applies to the measurement process. Typically, coastline length increases as increasing precision picks up smaller and smaller embayments; the process does not asymptote, as it would for a theoretically derived Euclidean curve, but continues to increase in an inverse logarithmic relationship with the length of the measure unit. What is striking is not that there is a relationship, but that it is as regular as it is. This is a manifestation of the fact that the tortuosity of the coastline seems to be scale-invariant; photographic enlargements of very small areas look, to the naked eye, like their larger siblings. Its ineluctable tortuosity on finer and finer scales confines the coastline to a nether-world intermediate between the smooth Euclidean curves of our schoolbooks and truly two-dimensional objects; hence, they have a fractional or *fractal* dimension intermediate between 1 and 2. This key attribute, self-similarity across a range of scales, is familiar to statistical physicists interested in critical phenomena, and also may be a property of many ecological processes.

The considerations raised by these examples are ones that must be addressed when we attempt to describe or simulate the dynamics of ecological systems. In particular, the problem of scale is one of the most basic we must face, since the description of pattern cannot be separated from the observational scale chosen. The hope must be that there are regularities underlying the different descriptions processes generate on different scales, and that those relationships are discoverable theoretically or by analysis of data. Without such regularities we can have no basis for extrapolating from one system to another, or one scale to another. Ecosystem experiments can be designed to elucidate these issues by providing information simultaneously on multiple scales. The rest of this short chapter is directed to these issues.

12.2 COMPLEX ECOLOGICAL SYSTEMS

In ecological systems, as elsewhere, complexity arises because processes interact on multiple scales. It has already been pointed out that different periodicities can interact to produce chaotic patterns. More generally, the interactions among scales—spatial, temporal, and organizational—and the need to study the behavior of systems across scales lend a complexity that is missed in studies that are restricted to a single scale. Any study is necessarily limited spatially and temporally, and in terms of the components that can be described. However, pattern is manifest on multiple scales simultaneously; and the investigator, by choice of perceptual scale, selects only a slice of the pie.

As suggested in the introduction, even the simplest nonlinear mathematical models describing population dynamics can exhibit very complex behavior (May, 1976). In particular, the discrete logistic,

$$N(t+1) = N(t) + rN(t)(1 - N(t)/K), \tag{1}$$

if initiated from positive population density $N(0)$, will always tend to the carrying capacity K when r is smaller than 2. However, as r is increased, the asymptotic dynamics pass through a sequence of bifurcations to cycles of length 2, 4, 8, 16, . . . , finally giving way to chaotic behavior when r is sufficiently large (May, 1974, 1976). The reason for this behavior is that the compensatory responses implicit in the nonlinear feedbacks are too strong, and cause the population density to overshoot the values that would lead to regulation.

Similar behavior can be observed, for analogous reasons, in continuous time models that include explicit delays, e.g. due to gestation, or even in models where the delay is implicit, as for example where the population is structured into age classes or stage classes. In the latter cases, since individuals must move through a series of explicit stages of development, a potentially stabilizing response that is mediated, for example, in the adult stage cannot be manifest immediately after the stimulus that initiated it; that is, the system exhibits a delay in its compensatory response. This delay is every bit as effective as an explicit one in leading to complex dynamics (Guckenheimer *et al.*, 1977).

Similar reasoning also can help us to understand the familiar oscillations observed in predator–prey systems, or at least in models of predator–prey systems, although these systems do not show chaotic oscillation. The rise of predator density in response to a rise in prey density is a potential regulator of the system. However, this regulatory response is delayed, because it takes time for the predator population to reach levels adequate to put a brake on prey densities; the same applies to the predator's negative response to reduced prey densities, and to the prey's response to increasing or decreasing predator densities. The delays prevent the system from achieving equilibrium, and result instead in the familiar limit cycle oscillations.

The problem of forced oscillations has already been mentioned. In systems where these are important, as for example when seasonal dynamics are superimposed on a system that has its own characteristic frequency of oscillation, the system may go through quite complicated fluctuations as the intrinsic and extrinsic periods compete for influence. One of the fundamental problems in examining data sets is to sort out the multiple causes of fluctuation. Methods such as Fourier analysis can be suggestive in identifying the primary modes of oscillation; but such methods lose their effectiveness in highly nonlinear systems in which oscillations are not merely additive.

The previous comments have been directed to the question of temporal fluctuations. Similar problems are faced in analyzing spatial patterns. In general, the key first step is to determine the degree to which the system is driven by external influences, and the degree to which its dynamics are autonomously controlled. This again is a property of scale; as dimension is increased, the system becomes more heterogeneous, less open to the external environment, and often statistically more regular. The questions of variability and predictability are intimately tied up with the question of scale, and cannot be addressed in the abstract.

The first observation is that the ecosystem or ecological community is operationally defined, according to the convenience of the investigator. The cell biologist has the luxury of knowing exactly what his or her fundamental unit is; the ecosystem scientist, on the other hand, must recognize that the measured properties of an ecosystem will vary as the scale and extent of description are changed. Different investigators, studying similar systems, are likely to view those systems through different prisms, because of the somewhat arbitrary definitions that must be made concerning the scale of description. For this reason alone it would be important to develop methods for studying how the dynamics of systems change across scales.

There are, however, other and equally compelling reasons. For the evolutionary ecologist it is clear that organisms perceive the environment through their own filters, and that the scales of variability experienced by diverse species, with distinct patterns of dispersal, dormancy, and other life history traits, will be equally diverse. In ecotoxicology the scaling problem is at the core of the science, whose validity rests in large part on the assumption that it is possible to scale up from laboratory bioassays and microcosms, and from field mesocosms to large-scale natural systems. In biotechnology, similar assumptions underlie the use of small-scale field testing of products that are intended for wide-scale applications.

As our awareness of global climate change and its consequences becomes more acute, so too does the desire in many quarters to couple the predictions of climate change with predictions of impacts on ecological and agricultural systems, and of their mutual effects on one another. But our understanding of these processes is on very different scales; the finest general circulation models, which form the basis for climate change prediction, use a basic grid cell measuring hundreds of miles on a side, whereas ecological models live in a far smaller world. The orders of magnitude difference between the scales of climate and ecological models has turned the attention of many to the need to develop methods to allow us to extrapolate across scales. Similarly, but to a lesser degree, the resolution limits associated with remotely sensed information mean that there are scale differences between the information that comes from those studies, and our understanding of the processes underlying it. The need to relate pattern on one scale to underlying processes on finer scales is a familiar one throughout science, and is the basis for extrapolation beyond the realm of experience.

12.3 SPATIAL AND SPATIOTEMPORAL PATTERN

In the oceanographic literature, the problem of patchiness and spatial heterogeneity in the distribution of planktonic species has been a topic of considerable attention since the middle of this century (Kierstead and Slobodkin, 1953; Steele, 1978). Information is collected routinely from ships navigating spatial transects through regions of interest, and analyzed to provide information regarding the spatial distributions of physical and biological variables.

These data are complemented by information derived from remote sensing. Standard statistical methods, such as Fourier analysis, are applied to determine the scales of aggregation and distribution, and the correlations among variables of interest. These are compared with the predictions from a variety of theories explaining patchiness, as the beginnings of a basic theory (Kierstead and Slobodkin, 1953; Platt and Denman, 1975; Levin *et al.*, 1989b; Morin *et al.*, 1989).

In terrestrial environments such approaches are less common among ecologists; in contrast, in the soils literature, there is a sophisticated and well-developed body of work on spatial statistics (Bras and Rodriquez-Iturbé, 1985; Burrough, 1983a,b). Methods such as Fourier analysis, kriging, semi-variograms, and other approaches for dealing with the interrelated problems of scale and measurement are central to the field, and to the related subject of geographical information systems (Ripley, 1981; Burrough, 1986). In vegetation studies, Kershaw (1973), Greig-Smith (1983), Oosting (1956) and others recognized the importance of stratified random sampling and other techniques to deal with scale, and R.H. Whittaker led plant synecology into the development of methods for describing the spatial distributions of species along gradients. The latter methods were primarily one-dimensional, however; and despite Whittaker's fascination with the importance of mosaic phenomena (Whittaker and Levin, 1977), he did not have the same success in dealing with multi-dimensional patterns as with one-dimensional distributions. In general, most of the approaches in the terrestrial environment have been static and descriptive, without adequate relation to mechanistic and dynamic theories. On the other hand, the large theoretical literature dealing with the dynamics of spatial pattern (e.g. Levin, 1976), has not made much of an attempt to compare its predictions with the static descriptions.

Now, in part because of vastly improved high-speed computers and advances in parallel processing, and in part because of advances in remote sensing, there exists tremendous potential to build bridges between theoretical investigations and empirical analyses. Spatial statistics, which represents an active area of theoretical investigation on its own, can and must be applied to describe the distributions of important physical, chemical, climatic, and edaphic variables. They must be coupled with similar analyses of biological variables to produce correlations that are at least suggestive of mechanistic interrelationships, and that provide tests of mechanistic models. Such mechanistic models are essential to making predictions beyond the range of our experiences, because reliance on correlations and extrapolations based solely on them can mislead badly (Lehman, 1986).

Our own approach (see Levin, 1989; Levin *et al.*, 1989a,b) has been to proceed simultaneously on multiple levels, developing suites of models of increasing complexity. These provide a bridge between complex data sets, and oversimplified caricatures that capture the essential features of data sets.

Such approaches have proved immensely useful in other applications in making clear what the basic reasons for observed patterns are (e.g., Ludwig *et al.*, 1978), and include Lorenz's powerful analysis of the problems of weather prediction, described earlier in this paper.

As a case in point, Kirk Moloney, Linda Buttel, and I have been interested in the gap phase patterns typical of many forests, grasslands, and intertidal habitats. These systems have the common feature that localized disturbances enhance regional coexistence by providing new opportunities for colonization, thereby creating a spatiotemporal mosaic from a system that at equilibrium would otherwise be low in diversity. In the serpentine grassland, where disturbances are the work of gophers and ants, we have joined forces with Harold Mooney and his research group to model the Jasper Ridge community. For forests, we are using a variety of local growth simulators, and will relate to data from both terrestrial and tropical forests.

Our approach is threefold. First, from remote sensing and ground surveys in the serpentine, the spatiotemporal dynamics of the plant species will be described, and spatial statistics of a variety of sorts applied. Simultaneously, a computer model has been developed that couples local growth and competition dynamics with the forces of disturbance and dispersal (Levin, 1989; Levin *et al.*, 1989b). Random disturbances of a variety of sizes reset the successional clocks in local environments, and open the way for recolonization. Local environments are spatially correlated with one another due to the spatial correlation in disturbance patterns, the local nature of propagule dispersal, and (in the more complicated versions of the model) neighbor competition. Such correlations can be represented through spatial correlograms, or equivalently, by semivariograms (e.g., Bras and Rodriquez-Iturbé, 1985). In the semivariogram the mean squared difference between the measurement of a variable at two spatially distinct points is given in relation to the spatial distance or lag between those points. Thus, the curve flattens out at large distances, when it exceeds the correlation length introduced by the already mentioned processes. Such relationships are observed in model results (e.g. Levin *et al.*, 1989a) or typically in data such as those reported by Robertson *et al.* (1988) in their interesting analyses of old field data (Fig. 12.1).

Another way to represent the patterns produced by such models, and to compare them with data, is to use a stratified technique in which the spatial or temporal variance of a measure is related to the size of the sampling window. For a system without underlying patterns of spatial heterogeneity, in which the only variability arises due to random local events, one expects variance to fall off with the size of the window. This is indeed the case; what is startling, however, is that for the simplest models we considered, and for scales on the order of the correlation length, there was a remarkable fit to a power law. This can be seen in Figure 12.2, in which log variability is plotted against log

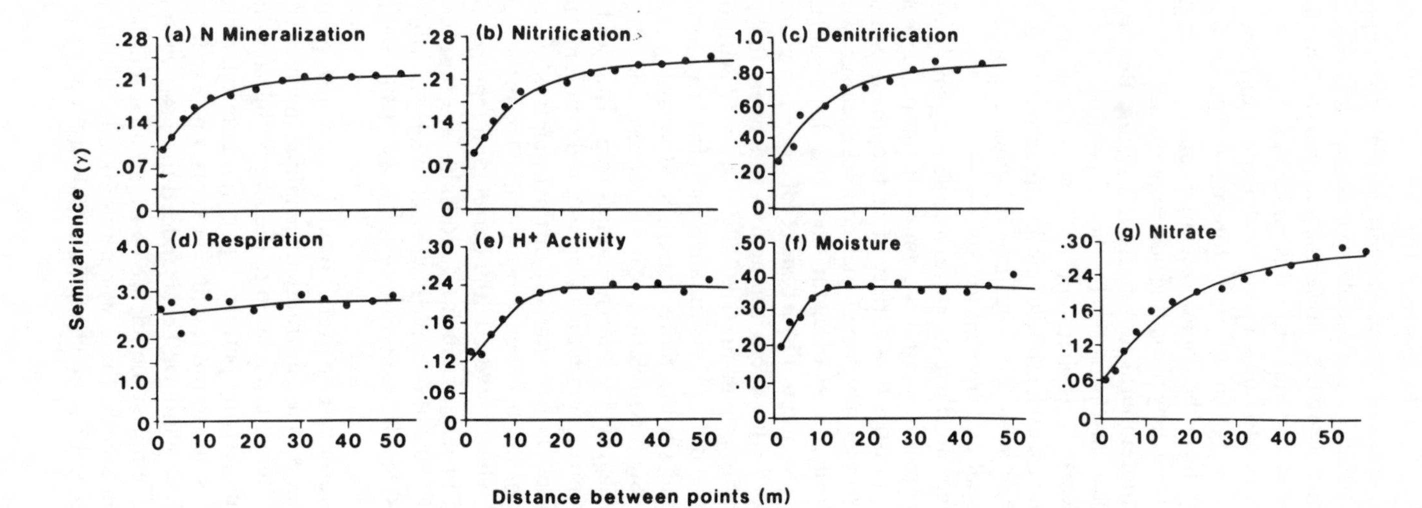

Figure 12.1 Semivariograms for (**a**) N mineralization potentials, (**b**) nitrification potentials, (**c**) denitrification, (**d**) soil respiration, (**e**) $[H^+]$, (**f**) soil moisture, and (**g**) soil nitrate across a 0.5-ha study site. Solid lines are best-fit models (reproduced with permission from Robertson *et al.*, 1988).

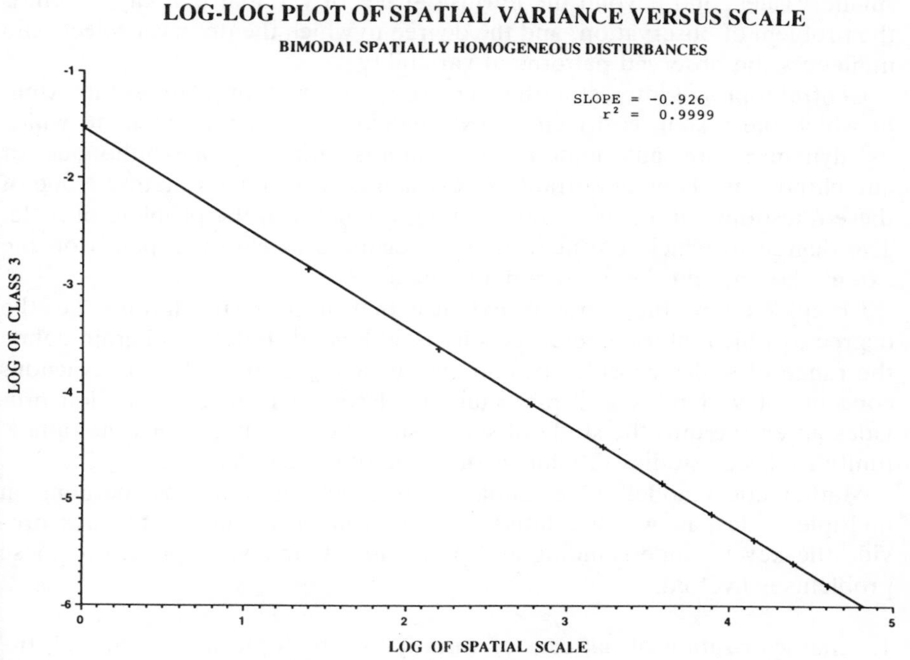

Figure 12.2 Model results: log of variance of abundance of successional class 3, versus log of size of basic observational window (see Levin *et al.*, 1989a).

of scale (window size). This phenomenon is reminiscent of similar relationships seen in statistical physics in the study of critical phenomena, and this similarity has led us to a collaboration with Richard Durrett at Cornell, with the intent to develop simpler versions of our model that still capture the main features, reproducing the observed scaling relationships. In this way we hope to understand the reasons underlying the pattern seen in Figure 12.2. By simulations and such analyses we endeavor to relate the critical exponents from the power law to statistics of disturbance and dispersal, and to provide a basis for extrapolation.

12.4 CONCLUSIONS

Multiple scale problems arise in the consideration of any ecological system, and are the essence of complexity. The description of pattern at one scale motivates the search for underlying mechanisms at finer scales, and for ramifications at broader scales. Effects also can cascade down from larger to smaller scales; and beyond the intrinsic scales of variability in any system is the problem of observation, and the degree to which the observer selects and influences the observed patterns of variability.

Central topics of interest in the study of any ecosystem relate to the extent to which the system is driven by external forces, and the extent to which its dynamics are autonomous. Are inputs primarily allochthonous or autochthonous? How important are exchanges with other systems? None of these questions can be addressed without attention to the problem of scale. The degree to which a system may be regarded as closed depends on the extent chosen, and the operational boundaries.

Complementing the notion of extent is that of grain (resolution), i.e. the degree to which internal heterogeneity is addressed. Extent and grain define the range of scales considered, and give meaning to the otherwise vacuous concepts of variability and predictability. Ecosystem experimentation provides an entrée into the study of such issues, because it gives us the opportunity to design studies to address the issue of scale explicitly.

Mathematical models are essential in any attacks on problems operating on multiple scales, in which relative rates of change at different scales provide the key to understanding and prediction. Our own approach to these problems is fivefold:

1. Characterization of patterns of variation on multiple scales, through the use of statistical methods such as spectral analysis.
2. Correlation of patterns of variation of potentially interdependent variables, and relation to known forcing functions.
3. Development of mechanistic models to explain observed patterns, to allow discrimination among competing hypotheses, and to provide a basis for extrapolation.
4. Exploration of simplified models, in order to abstract the essential features underlying observed behaviors.
5. Closing of the loop by coupling model development and exploration with experimentation, and by comparison of model outputs with field observations from experimental systems.

In general, different mechanisms will be seen to be responsible for observed patterns on different scales; hence a suite of models are needed, emphasizing different processes on different scales. For example, in our studies of the distribution of krill in the Southern Ocean (Levin *et al.*, 1989b), we have concluded that physical circulation models explain the distribution of organisms on the larger spatial scales, through their movement of large patches of water. On smaller scales the explanatory power of the physically based models breaks down completely, however, and detailed models of the swimming and aggregation behavior of the euphausids must be invoked.

Coupled with these distinct approaches at different scales must be studies of how the effects of particular processes are translated across scales, and the degree to which scale invariants such as those involved in self-similar phenomena provide a basis for extrapolation.

The study of complex systems is the study of the interaction between processes acting on multiple scales. Such problems are among the most exciting in ecology today, and are central to addressing phenomena associated with global change. New and innovative mathematical approaches will be needed to deal with these problems, coupling advances in dynamical systems theory, in stochastic processes, in statistics, and in supercomputing. As has been true in the past, ecology not only will benefit from these approaches, but will stimulate developments in mathematics, statistics, and computing, bringing vitality to all of these subjects through mutual amplification of ideas and methods.

ACKNOWLEDGEMENTS

Support by National Science Foundation grant BSR-8806202 to the author is gratefully acknowledged. This publication is ERC-206 of the Ecosystems Research Center at Cornell University, and was supported in part by the US Environmental Protection Agency Cooperative Agreement CR812685 with Cornell University. The work and conclusions published herein represent the views of the authors, and do not necessarily represent the opinions, policies, or recommendations of the funding agencies.

REFERENCES

Bras, R.L. and Rodriquez-Iturbé, I. (1985) *Random Functions and Hydrology*, Addison-Wesley, Reading, MA.

Burrough, P.A. (1983a) Multiscale sources of spatial variation in soil. I. The application of fractal concepts to nested levels of soil variation. *J. Soil Sci.* **34**, 577–97.

Burrough, P.A. (1983b) Multiscale sources of spatial variation in soil. II. A non-Brownian fractal model and its application in soil survey. *J. Soil Sci.* **34**, 599–620.

Burrough, P.A. (1986) *Principles of Geographical Information Systems for Land Resources Assessment*, Clarendon Press, Oxford.

Gleick, J. (1988) *Chaos: Making a New Science*, Penguin Books, New York.

Greig-Smith, P. (1983) *Quantitative Plant Ecology*, Butterworth, London.

Guckenheimer, J., Oster, G. and Ipaktchi, A. (1977) Dynamics of density dependent population models. *J. Math. Biol.* **4**, 101–47.

Kershaw, K.A. (1973) *Quantitative and Dynamic Plant Ecology*, 2nd edn, Arnold, London.

Kierstead, H. and Slobodkin, L.B. (1953) The size of water masses containing plankton blooms. *J. Mar. Res.* **12**, 141–7.

Lehman, J.T. (1986) The goal of understanding in limnology. *Limnol. Oceanogr.* **31**, 1160–6.

Levin, S.A. (1976) Population dynamic models in heterogeneous environments. *Ann. Rev. Ecol. System.* **7**, 287–311.

Levin, S.A. (1990) Physical and biological scales and the modelling of perdator–prey interactions in large marine ecosystems. In: Sherman, K. and Alexander, L.M. and Gold, B.D. (Eds), *Large Marine Ecosystems—Patterns, Processes, and Yields,* AAAS Selected Symposium, American Association for the Advancement of Science Publ. No. 90–305, Washington, DC.

Levin, S.A., Moloney, K., Buttel, L. and Castillo–Chavez, C. (1989a) Dynamical models of ecosystems and epidemics. *Future Generation Computer Systems.* **5**, 265–74.

Levin, S.A., Morin, A. and Powell, T. (1989b) Patterns and processes in the distribution and dynamics of Antarctic krill. In: *Conservation of Antarctic Marine Living Resources.* (CAMLR) Selected Scientific Papers, Part 1, 281–299, Hobart, Tasmania.

Lorenz, E. (1963) Deterministic nonperiodic flow. *J. Atmos. Sci.* **20**, 448–64.

Lorenz, E. (1964) The problem of deducing the climate from the governing equations. *Tellus* **16**, 1–11.

Ludwig, D., Jones, D.D. and Holling, C.S. (1978) Qualitative analysis of insect outbreak systems: the spruce budworm and forest. *J. Anim. Ecol.* **47**, 315–32.

Mandelbrot, B. (1977) *The Fractal Geometry of Nature*, W.H. Freeman, New York.

May, R.M. (1974) Biological populations with non-overlapping generations: stable points, stable cycles, and chaos. *Science* **186**, 645–7.

May, R.M. (1976) Simple mathematical models with very complicated dynamics. *Nature* **261**, 459–67.

Morin, A., Okubo, A. and Kawasaki, K. (1989) Acoustic data analysis and models of krill spatial distribution. Background paper for Annual Meeting of Commission for Conservation of Antarctic Marine Living Resources (CCAMLR), October 1988, Hobart, Australia.

Oosting, H.J. (1956) *The Study of Plant Communities*, 2nd edn, W.H. Freeman, San Francisco, CA.

Platt, T. and Denman, K.L. (1975) Spectral analysis in ecology. *Ann. Rev. Ecol. System.* **6**, 189–210.

Ripley. B.D. (1981) *Spatial Statistics*, John Wiley, New York.

Robertson, G.P., Huston, M.A., Evans, F.C. and Tiedje, J.M. (1988) Spatial variability in a successional plant community: patterns of nitrogen availability. *Ecology* **69**, 1517–24.

Ruelle, D. and Takens, F. (1971) On the nature of turbulence. *Commun. Math. Phys.* **20**, 167–92.

Schaffer, W.M. and Kot, M. (1985) Nearly one-dimensional dynamics in an epidemic. *J. Theor. Biol.* **112**, 403–27.

Schaffer, W.M., Ellner, S. and Kot, M. (1985) Effects of noise on some dynamical models in ecology. *J. Math. Biol.* **24**(5), 479–523.

Steele, J. (1978) Some comments on plankton patches. In: Steele, J. (Ed.), *Spatial Pattern in Plankton Communities*. NATO Conference Series, Series IV: Marine Sciences, vol. 3. Plenum Press, New York, pp. 11–20.

Whittaker, R.H. and Levin, S.A. (1977) The role of mosaic phenomena in natural communities. *Theor. Pop. Biol.* **12**(2), 117–39.

Part 2

GROUP REPORTS

13 Aquatic Ecosystem Experiments in the Context of Global Climate Change: Working Group Report

J.F. KITCHELL (*Group Leader*), S.R. CARPENTER (*Rapporteur*), S.E. BAYLEY, K.C. EWELL, R.W. HOWARTH, S.W. NIXON, and D.W. SCHINDLER

13.1 INTRODUCTION

Water is a fundamental societal need. Global climate change will affect both the quantity and quality of water resources. Although the specific features of rates and magnitude remain the subject of continuing debate, all reasonable forecasts of global climatic response to increases in atmospheric carbon dioxide and other greenhouse gases include increases in mean temperatures at middle latitudes and a substantial increase in mean sea level (NRC, 1986). Coastal systems and the regional hydrologic cycles of continents will be profoundly altered. While there is no doubt that the riverine conduits and oceanic recipients of a changing hydrologic cycle will be responsive to global change, we have focused our attention on the systems we deem most sensitive and amenable to the use of whole-system experimentation. With these as givens, our group attempted a collective vision of an unknown future. We constructed a consensus of the ecosystem-scale experiments that would offer most powerful insights to interests of both the research and management communities.

Ecosystem experiments resolve controversy and rapidly achieve insights that would take far longer through observational and/or small-scale experimental studies (Likens, 1985). The limiting nutrient controversy in limnology is one example of the power of ecosystem experiments to reduce disagreement among scientists and prompt effective management action (Schindler, 1988). By 1970, several decades of debate deriving from correlational and small-scale experimental studies still could not resolve the relative importance of carbon, nitrogen, and phosphorus in controlling water quality of eutrophic lakes (Likens, 1972). Whole-lake enrichment experiments showed clearly

Ecosystem Experiments. Edited by H.A. Mooney *et al.*

that phosphorus was the critical nutrient. Phosphorus control is now the central element of water quality management (Schindler, 1988).

Proper scaling in space and time is an essential requisite in the design of ecosystem experiments. If the areal or temporal dimensions are insufficient, results are likely to be confusing or misleading (Carpenter and Kitchell, 1988). Statistical issues are subordinate to scale considerations as the history of statistics derives from subsampling and replicated test plots rather than the holistic view of entire systems. Although ecosystem experiments often will not meet the replication requirements of conventional statistics, the need for effective research and management at the whole-system level demands that we overcome the constraints of statistical tradition (Hurlbert, 1984). Fortunately, new statistical and analytical tools (Walters, 1986; Carpenter *et al.*, 1989) provide a rigorous foundation for detecting change in experimentally manipulated systems.

Ecosystem experiments remain rare despite their large, positive impact on the progress of science and development of management practices. Lack of access to research sites of sufficient size and institutional constraints are the main limitations. The prospects of global climate change demand that our society effectively plan for and respond to extraordinary changes in our water resources. We must accept the responsibility for extraordinary commitments of resources and research talent in order to gain the powerful insights of ecosystem experimentation.

In this report we emphasize scientific problems requiring the scale and understanding that can be obtained only by the power of whole-ecosystem experiments. We emphasize those types of systems and specific issues where we feel societal interests will be best served by immediate attention. Assessments and recommendations are presented for discrete system types but include the assumption that many important changes in aquatic systems will derive from changes in the terrestrial systems located 'upstream' in the hydrologic cycle. In addition, we point to certain major research questions that may be better suited to conventional methods such as comparative or survey approaches, incisive and well-replicated experimental designs in microcosm or mesocosm systems, process rate responses and modeling.

13.2 ASSESSMENT AND RECOMMENDATIONS

13.2.1 LAKES

Freshwater resources will be generally reduced by climate change. They will become increasingly contaminated by organic compounds, toxic metals and nutrients as the result of land-use changes and long-range transport in the atmosphere. Freshwaters are already in short supply in many areas of the world. Research on the effects of global change on freshwaters must therefore be considered to be of the highest priority.

As abiotic conditions (e.g. mean and maximum temperatures) change with climate shifts, major biogeographic changes will occur. Differences in dispersal capacity, increased variability in local populations, local extirpations, and invasion by exotics will occur. Species of primary interest as resources (e.g. fishes) will respond in ways that create important challenges to management.

In an ecosystem context the most important changes to consider are temperature effects, hydrologic inputs (flushing rate and nutrient load) and contaminant load. New whole-lake manipulations are needed in two major areas.

Contaminants

As freshwater resources become scarce, the problem of understanding and controlling contaminant dynamics will become increasingly important. Both organic and metal contaminants need to be considered. Our ignorance of contaminants has led to many global problems. For example, revolatilization of PCBs from the Great Lakes is now spreading these contaminants to other ecosystems. Mammals and fishes from the high arctic are known to be contaminated with toxaphene, dieldrin, and other organic compounds that have been released at temperate latitudes. Acidification of lakes is thought to be responsible for the increased mercury in fishes, although the mechanism is not known. Whole-lake experiments are already under way with cadmium, which is approaching toxic levels in Lake Michigan and other lakes. New experiments on organochlorine pollutants, other organic compounds, and several trace metals are needed. In anticipation of global climate change, these experiments must emphasize synergistic interactions including; temperature × contaminant, food web structure × contaminant, and acidification × metal. The mechanistic understanding derived from these experiments will substantially enhance our predictive strength in dealing with similar issues in coastal systems where whole-system experimentation is more difficult (see 'Flooding of coastal wetlands' below).

Temperature

As mentioned above, changing climate scenarios all include a warming trend in those regions of the northern hemisphere that contain the majority of the world's lakes. We can logically expect that the thermal limits of current biogeography will change. Studies of ponds and lakes heated by effluents from power plants have provided some information about the effects of temperature change on aquatic ecosystems. However, no study has included adequate premanipulation data or good reference ecosystem studies. A thorough study of thermal effects on lakes and streams is required, and might be relatively inexpensive if combined with construction of new power plants.

Some of the major changes anticipated in aquatic systems do not require whole-lake experiments; many major effects on freshwater lakes are predictable from current knowledge. For example, reduced water flows will cause increased concentrations of all chemicals including both nutrients and contaminants. These are predictable from models developed for eutrophication. Salinization will be a growing problem for shallow lakes in increasingly arid areas, but whole-lake manipulations are not needed for this issue. Preservation or replacement of fisheries will present major challenges to managers. For example even moderate warming may eliminate some major coldwater fisheries, requiring that they be replaced by warmwater assemblages. Fisheries and aquatic community changes will require much research, as well as development of new management capacities, but not necessarily whole-lake manipulations.

13.3.3 COASTAL AND WETLAND SYSTEMS

Changes in hydrologic fluxes and in sea level will cause enormous changes in the global extent and nature of wetland and estuarine systems. Coastal systems will experience the dual pressures of development and increased sea level. Much of the world's wetland area lies in permafrost regions between 50 and 70 degrees north latitude, and will obviously be subject to dramatic changes associated with global warming. Two major aspects of global climate effects on wetlands must be considered. (1) Wetlands are major sources of methane, one of the more potent greenhouse gases. Changes in the extent and function of wetland ecosystems could have significant feedback on climate. (2) Coastal wetlands are the interface between terrestrial and aquatic systems. Their role as sinks or sources of nutrients and contaminants will be of substantial local and regional concern. While many of the concerns about changes in coastal and wetland systems can be anticipated by specific experiments (e.g. thermal effects, increased flooding) and more accurate measurements (e.g. gas flux), certain issues require the scale of whole system experiments. Among those are:

Warming of boreal wetlands

Permafrost (thin lenses of ice 1–2 m deep beneath 0.5 m of peat) underlies much of the vast boreal wetlands of the northern hemisphere. Global warming will result in major changes in the hydrology of these systems and their coupling to the atmosphere. Increased erosion and subsidence of frozen peat can cause major problems for extant and future transportation systems. We do not know if melting of permafrost will cause flooding or drainage of boreal wetlands, or how this will alter wetland growth and areal extent. We do not know if methane fluxes will increase or decrease with a change in water level. These changes cannot be forecast from small-scale studies; ecosystem experimentation is essential.

Flooding of coastal wetlands

Coastal wetlands such as salt marshes and mangrove swamps, plus their associated estuaries, serve as nurseries for the fertile rim of our world oceans that produces over 90% of total marine fisheries yield. The majority of the human population lives within a few tens of kilometers of those coasts. As a consequence, coastal wetlands have become major sinks for anthropogenic toxic compounds such as mercury, lead, and cadmium as well as a variety of organic pollutants such as PCBs and aromatic hydrocarbons. The amounts of toxic material in the sediments of coastal marshes and estuaries are truly staggering. Some salt marshes near major industrial areas contain the largest quantities of toxic pollutants found in any natural ecosystem.

Intertidal vegetation is adapted to fluctuating water levels and relatively insensitive to pollution. So long as the wetland systems are intact, pollutants are safely entombed with relatively little impact on estuarine and coastal ecosystems. In estuaries the initial mixing of fresh and salt waters induces flocculation and sedimentation of toxic-laden organic matter resulting in a similarly inactive archive for riverborne pollutants. These fortunate results of evolution and physics will fail as global climate changes.

The rate of sea level rise is predicted to accelerate as a result of global warming. Based on previous experiments and observations the expected rate of rise will occur more rapidly than the natural system can accommodate, resulting in increased anoxia in more continuously flooded sediments and the ultimate death of the primary vegetation. In the absence of sediment accretion and retention due to healthy plants, the toxic-laden materials now resident in coastal marshes will be mobilized in a massive series of erosional events. Similarly, changes in freshwater inputs and mean sea level will affect the previous and future depositional sites of estuarine sediments. While the redistribution of sediment particles is reasonably predictable, we do not know the fate of their toxic contaminant load and its potential effects on the coastal biotic resource system. Clearly, the complexity of prospects and value of resources in question demand the holistic view and insights afforded only through whole-system experimentation.

Study site availability is a major constraint in this case. Competing interests for coastal sites require that a concerted effort be mounted. Some of the most important questions are amenable to microcosm- and mesocosm-scale studies. If those can be intensified and properly coordinated with the development of an experimental estuarine reserve (modeled after Canada's Experimental Lakes Area), a maximum of research results might derive with minimal constraint. Freshwater ecology has experienced substantial gains from the capacity to manipulate whole lakes. The dimension of unknowns and societal value of coastal marshes and estuaries strongly reinforce the urgent need for attention to development of both the facilities and results of whole-system experimentation.

Draining/flooding of inland wetlands

As hydrologic changes occur, many inland wetlands will be gradually dried or increasingly flooded. As in the case of boreal peatlands, the net effects on gas flux have major portent for atmospheric feedback. As in the case of coastal marshes, nutrient and contaminant fluxes will be fundamentally different and are not easily predicted. In addition, changes in biological communities will evoke need for attention to management alternatives appropriate for protection and maintenance of desired resources. Again, the net effect of changing hydroperiod cannot be simulated in mesocosms and requires experimentation at the ecosystem scale.

13.3 CONCLUDING REMARKS

Many readers of this report will find it insufficient. We have admittedly centered our attention on a small subset of the issues and ecosystems that might feel the effects of a rapidly changing climate. Our goal has been to identify those issues and places most important to a society that rightfully expects good guidance from its scientific community. The logical consequence of global climate change is a reallocation of water and the resources it creates. We cannot change that. We can, however, offer specific advice about ways to anticipate and respond to some of the ancillary problems that may develop. Using the power of ecosystem experimentation as the most appropriate tool, we emphasize research on the following prospects:

1. As regional warming occurs, biological communities will disassemble through differential deletions and invasions. New communities will emerge. New management practices will be necessary. Whole-system thermal enhancements can serve to anticipate those responses.
2. Regional warming will profoundly affect wetland systems. In addition to the obvious physical and biological concerns, the prospect of a major change in flux of greenhouse gases must be expected and determined. The possible feedback to atmospheric systems requires large-scale evaluation.

3. Changes in regional hydrology and mean sea level create an important and ironic prospect: sites containing the buried wastes from our past may be disturbed and their toxic contents mobilized to the extent that we must be concerned about major effects in the future. A system of experimental ecosystems in the coastal environment can help guide responses to that prospect.

As more effective forecasts of global climate change emerge, the relative importance of these issues will change. Regardless of that, we encourage the development of perspective and institutional commitments required to increase the use of ecosystem experimentation as a research tool.

REFERENCES

Carpenter, S.R. and Kitchell, J.F. (1988) Consumer control of lake productivity. *BioScience* **38**, 764–9.

Carpenter, S.R., Frost, T.M., Heisey, D. and Kratz, T. (1989) Randomized intervention analysis and the interpretation of whole ecosystem experiments. *Ecology* **70**,142–52.

Hurlbert, S.H. (1984) Pseudoreplication and the design of ecological field experiments. *Ecol. Monogr.* **54**, 187–211.

Likens, G.E. (Ed.) (1972) *Nutrients and Eutrophication: the Limiting Nutrient Controversy*, Am. Soc. Limnol. Oceanogr. Spec. Symp. 1.

Likens, G.E. (1985) An experimental approach for the study of ecosystems. *J. Ecol.* **73**, 381–96.

National Research Council (1986) *Global Change in the Geosphere–Biosphere*, National Academic Press, Washington, DC.

Schindler, D.W. (1988) Experimental studies of chemical stressors on whole lake ecosystems. *Verh. Internat. Verein. Limnol.* **23**, 11–41.

Walters, C.J. (1986) *Adaptive Management of Renewable Resources*, Macmillan, New York.

14 Whole-terrestrial Ecosystem Experiments in the Context of Global Change: Working Group Report

B.H. WALKER (*Group Leader*)

Discussion Participants
P. ATTIWILL, D. GRAETZ, J. MacMAHON, E. MEDINA, H. MOONEY, E.-D. SCHULZE, C.O. TAMM, F. TRILLMICH, R. WRIGHT

14.1 INTRODUCTION

Four main kinds of global changes are occurring that impact terrestrial ecosystems of the world:

1. The atmosphere of the world is changing, chiefly with increases in CO_2, NO_x, CH_4, and other trace gases. The direct effects of these gases, both singly and in combination, are fertilizing (CO_2, N), acidifying and otherwise altering the atmospheric and soil chemical environment of plants. A special aspect of this kind of change is the decrease in stratospheric ozone which, if it occurs significantly in middle and lower latitudes, could have grave consequences on biotic systems in terms of UV radiation effects.
2. As a result of the increase in CO_2, CH_4, and other 'greenhouse' gases, the temperature, and consequently the climate, of the earth is likely to change more rapidly than it has ever changed in the past. There are some indications that it is already changing, but the inherently large temporal and spatial variability in climate masks any clear indications of early changes. Although current predictive global circulation models are still under development, there is an emerging consensus from them that the trend will be to higher temperatures, with the effect increasing towards the poles, with generally higher and probably more variable rainfalls.
3. Land-use patterns over the globe have already had a very significant effect on regional hydrology and soil movement, and on albedo, and the effects

Ecosystem Experiments. Edited by H.A. Mooney *et al*.

are intensifying as land 'development' extends into the tropics and semi-arid regions of the world.

4. The biodiversity of the world is declining. The factors contributing to this are species extinction (complete loss), local extinction (or local loss of species), and a reduction in evenness (i.e. although present, many species are now rare and no longer play a significant ecological role in the community). The combined consequences of these declines for ecosystem functioning are not known, although they have been considered theoretically.

Given the uncertainties in predicting these changes at local and regional scales, and given our present inadequate knowledge of the ways in which species and ecosystems will respond to such changes, it is very unlikely that we will be able to develop ecosystem models, based on all the processes involved, that can be used reliably in the short term. Accordingly, the goal of this group was to consider the above changes and to identify those questions that warrant priority attention in terms of whole-ecosystem experiments; both because of their importance and because the questions concerned cannot be satisfactorily answered in any other way.

This group concluded that:

1. The consequences of widespread land-use changes would be best approached through well-planned observations and meaurements of existing situations, coupled to the development of conceptual and, subsequently, quantitative models.
2. The global-scale consequences for ecosystem function of decreasing biodiversity are as yet insufficiently understood to allow for (or justify) the scale and cost of a direct large-scale experimental approach. There is a preliminary requirement for a synthesis of the experimental and other empirical data available from local, fine-scale effects in order to develop sound, testable predictions of the effects (if any) at global scales.
3. Our understanding of the changes in stratospheric ozone, and the levels of change likely to have serious consequences, is at this stage still too poor to warrant a direct experimental approach at the level of a whole ecosystem.
4. Two major questions emerge, which could and should be approached through whole-ecosystem experimentation:
   (a) By how much will increasing atmospheric CO_2 and/or other trace gases affect net primary production in natural ecosystems, and by how much will it modify carbon storage in soils? How will changes in nutrient availability (both atmospheric and in soil) influence these effects?
   (b) How will the effects of a change in climate—primarily changes in precipitation and temperature—on ecosystem composition and production be modified by increasing atmospheric CO_2?

These two questions have also been identified by the International Geosphere Biosphere Program's Co-ordinating Panel on Effects of Climate and Atmospheric Change on Terrestrial Ecosystems, as being of the highest priority.

14.2 PROPOSED EXPERIMENT

14.2.1 PRELIMINARY CONSIDERATIONS

The experiment that is proposed is directed to answering the above two questions. It will involve manipulating levels of CO_2 in combination with water and/or temperature, in selected ecosystems, and measuring the responses of these systems over time. It will clearly be a major effort, involving considerable expense and research investment, and many choices will have to be made in terms of the exact nature of the treatments (levels, transient vs single increments, variables to be measured, etc.). Its design and the questions it seeks to answer, therefore, must be rigorously developed. In this section we restrict consideration to the main features of the experiment. The more detailed aspects of the available experimental procedures and techniques are dealt with in the next chapter.

The experiment, or experiments, will constitute part of a large international effort aimed at a predictive understanding of global change, and in relation to this larger effort, early consideration needs to be given to:

1. What information, relevant to the two questions, can be obtained through observations of existing gradients, changes, etc. that will: (a) keep the requirements for the experiment to a minimum, and (b) complement the results in a way that allows their extrapolation to larger time and space scales.
2. Collaboration with others (particularly general atmospheric circulation and global vegetation modellers) who will use the results. In addition to their inherent value, for a variety of purposes, efforts to devise mechanistic models of global vegetation change (under the auspices of the IGBP) have already identified the second question as a major constraint to the development of such models. To ensure that maximum use is made of the experimental results, early collaboration with other potential users of the results is important.

The two questions identified as being of primary concern have sufficient overlap to suggest that a combined experimental approach is possible. Both involve changes in CO_2. The first includes the interactive effects of nutrients, and the second includes the effects of water and temperature. There are indications that, for many biomes, nutrients may limit or moderate the photosynthetic response to CO_2 enrichment. Also it can be argued that CO_2 effects on vegetation will be appreciable only in the absence of climate change because the climatic factors of precipitation and temperature, in particular, are far more influential in determining plant productivity and demography than CO_2. These considerations call for experiments with combined CO_2, precipitation and temperature treatments.

14.2.2 DESIGN CONSIDERATIONS

Ecosystem selection

Sensitivity to global change, in terms of the two questions being posed, will be a function of the expected amounts of change in rainfall and temperature, and the potential of the system to change. Two ecosystem attributes which will strongly determine the ways in which ecosystems will both respond to atmospheric and climatic change, and feedback on atmospheric change, are their present rates of gross primary production (or turnover), and their amounts of stored carbon. Therefore, in order to develop the capacity for global predictions through interpolative modelling, the experiment needs to be conducted under conditions of high and low ecosystem productivity, and high and low soil carbon storage. Examples of the four combinations are shown in Table 14.1.

Stored soil carbon is primarily a function of temperature. Response to improved (as opposed to merely increased) soil moisture conditions will depend strongly on soil nutrient status and temperature, and response to increasing temperatures will depend on the extent to which low temperatures and soil moisture are presently limiting factors. Figure 15.1 (in the following chapter) places the major biomes of the world in the available moisture/available nutrients plane.

Combining the considerations of carbon storage, present gross productivity, rainfall, and soil nutrients, the biomes most suited for study are:

1. boreal forest or tundra
2. tropical rainforest

Table 14.1

	Low production	High production or turnover
Low carbon storage	Deserts or arid savannas	Tropical rainforests
High carbon storage	Tundra or boreal forest	Temperate forests (deciduous or coniferous) or temperate grasslands

3. arid savanna or shrubland
4. temperate forest or grassland.

Site selection

Within the chosen biomes the selection of sites will clearly be difficult. The factors to be considered are:

1. soil type and chemistry: widespread and typical of the biome;
2. size: internally homogeneous, large enough to eliminate edge effects;
3. vegetation state: in each case the history of the site must be carefully considered in relation to the disturbance regime of the biome and community type concerned, and the selected site should be modal in this regard;
4. air pollution: the sites must be selected both in areas free of significant air pollution, now and during the lifetime of the experiments, and in areas currently influenced by pollution (particularly atmospheric N); the results of the experiment must apply to both kinds of areas, and the effects of CO_2 could well be different.

Experimental design

Consideration must be given to the use of free air carbon dioxide enrichment (FACE) to avoid the confounding effects of closed chambers on air movements, temperature, and humidity gradients, etc. (see the following chapter). It may be useful, however, to explore certain questions, in a preliminary way, using large enclosed plant communities. The potential availability of the enclosed Norwegian small forest catchments (see Chapter 9 in this volume) for this purpose should be explored.

It is proposed that the main focus of the study will be the field-scale, manipulative experiment involving increased CO_2 under free air circulation (FACE) wherever feasible. However, understanding the results, and incorporating them into subsequent mechanistic models, will require a series of closely integrated, controlled experiments across a range of scales, from single leaves in cuvettes through whole plants and enclosed small plots to the large open trial. This approach is described in more detail in Chapter 15, and is also recommended by the IGBP working group that considered this problem (see IGBP Report 11, Section II).

The ideal of a well-replicated, full factorial experiment with randomized interspersion of treatments will be extremely difficult, but is necessary for a full test of the hypotheses. As a preliminary suggestion, the minimum necessary treatments are:

1. control;
2. CO_2 increase—doubling the 1980 concentration;
3. N increase (where appropriate)—provided as NH_4NO_3, at two levels;
4. water increase—50% increase with both similar and different seasonality and storm size distribution;
5. $CO_2 \times N$ (where appropriate);
6. $CO_2 \times$ water.

Technologies available for field experiments with changes in gaseous levels are given in the following chapter.

Modifying temperature in a controlled manner presents a major difficulty. If it is practically feasible to do so, then the extra treatments required would be: temperature increase, $T \times CO_2$, $T \times H_2O$, $T \times H_2O \times CO_2$. The final selection of treatment combinations should be based on a carefully determined set of questions, specific for each site. In forests each treatment will require at least 1 ha, and considering the numbers of treatments (*c.*8) and replicates (5?), the size of the experiment will be over 40 ha.

Variables and processes

In determining the response to treatments as many aspects as possible should be taken into account, involving a wide array of scientists. The full range of species types should be examined (plants, vertebrates, invertebrates, microorganisms), and the processes should include both ecosystem level processes (NPP, E/T, decomposition) and demographic processes (seedling germination and establishment, age-specific mortality, relative growth rates, reproduction).

Secondary effects and episodic events

In many ecosystems, functional properties such as NPP are determined by the mean values of temperature, soil water availability, etc., but demography is determined by rare events (occasional very wet years, droughts, aseasonal frosts, hurricanes) and secondary effects, such as fire (which is a function of productivity, accumulated fuel, and seasonality). These features cannot be included in the experimental design, though they may enter into the experiment by chance. They must, however, be taken into account in extending the results of the experiments through modelling. This leads on to a general consideration of time and space problems.

14.2.3 TIME AND SPACE SCALES

Changes of parameters on a global scale are relatively slow compared with rates of ecosystem processes. One reason for this lag is that events occurring in one place must have their effects move through space, and this takes time. The problems of the space–time continuum and the attendant problems of scaling are inherent to ecological research.

In the research that we propose, the constraints of the space–time continuum force us to ignore certain fundamental alterations in ecosystems that are occurring in response to global change in climates, land-use patterns, pollution effects, etc. Our experimental studies will be most significant for processes that occur within short periods. We assume that we can extrapolate to longer periods, but this cannot be known with certainty. Our own experimental protocols generally propose measuring events that happen rapidly, in the sense of ecological time. The measurements we make, and the processes that we choose to study, are, however, thought to be good 'benchmarks' that can be used to indicate longer-term changes.

One aspect of the experiments that will need to be measured, and included in the analysis and interpretation of all results, is that as we experimentally alter ecosystems, various components of these systems will be changing at different rates. For example, during the time we measure a change in net primary production that might be induced by altered CO_2 levels, the entire microbial flora may have changed its species or functional composition to the point where it is no longer the same community. In contrast, long-lived trees may persist despite the fact that many other ecosystem components have been replaced.

It should be obvious that if different processes and components of the same ecosystem differ in their response times, then different ecosystems might also differ. That we realize this at the outset may permit us to compensate by very carefully devising our experiments and interpreting the results.

15 Available Technologies for Field Experimentation with Elevated CO_2 in Global Change Research

B.R. STRAIN (*Group Leader*)

Discussion Participants
L.H. ALLEN, Jr, D. BALDOCCHI, F. BAZZAZ, J. BURKE, R. DAHLMAN, T. DENMEAD, G. HENDREY, A. McLEOD, J. MELILLO, W. OECHEL, P. RISSER, H. ROGERS, J. ROZEMA and R. WRIGHT

15.1 INTRODUCTION

Technology is available, or envisioned, to accomplish experimental manipulation of atmospheric factors in ecosystems representative of global biomes. In some cases, however, it will be necessary to modify and scale up existing equipment originally designed and fabricated for use on smaller samples. Ecosystem environmental control facilities must be of adequate size to contain the ecosystem sample desired. The primary problem is control of atmospheric gases, particularly CO_2 and water vapor. It is, of course, also necessary to control air temperature and the energy and hydrological balances in studies of global change.

It must be recognized that it is not possible to obtain ideal simulation of the atmospheric environments of open ecosystems. Rather, the goal is to apply the best technology and treatment scenarios available to obtain a degree of control sufficient to interpret the potential effects of global change.

Ecosystem site selection will be dependent upon the availability of adequate technology to reach desired levels of control for each system. Only a few ecosystems have been studied and only three have been experimentally manipulated for the CO_2 variable. The only systems analyzed to date are all graminoid in structure. Oechel and colleagues (Oechel and Strain, 1985) enriched tundra plots for three years (1983–85), Drake *et al.* (1989) completed four years of treatment of representative plots of estuarian salt marsh (1987–89), and Owensby controlled tall-grass prairie at his site in Kansas in

Ecosystem Experiments. Edited by H.A. Mooney *et al.*

1989 and 1990. Some autecological studies have been accomplished in the other ecosystems shown in cross-hatching in Figure 15.1, but system-level data exist only for graminoid communities. It will be necessary to obtain information on more representative systems from the matrix shown in Figure 15.1 if we are to develop a predictive capability for the effects of long-term global change on terrestrial ecosystems.

When ecosystems are chosen for large-scale study of global change, they should represent the range of diversity indicated in Figure 15.1. It is possible to classify broadly the world's ecosystems into six biomes. Ranked roughly in order of predicted sensitivity to the Manabe Scenario (Manabe and Wetherald, 1980) of global change (CO_2 increase, global warming with increasing temperature from equator to high latitudes, mid-latitude drying) they are: (1)

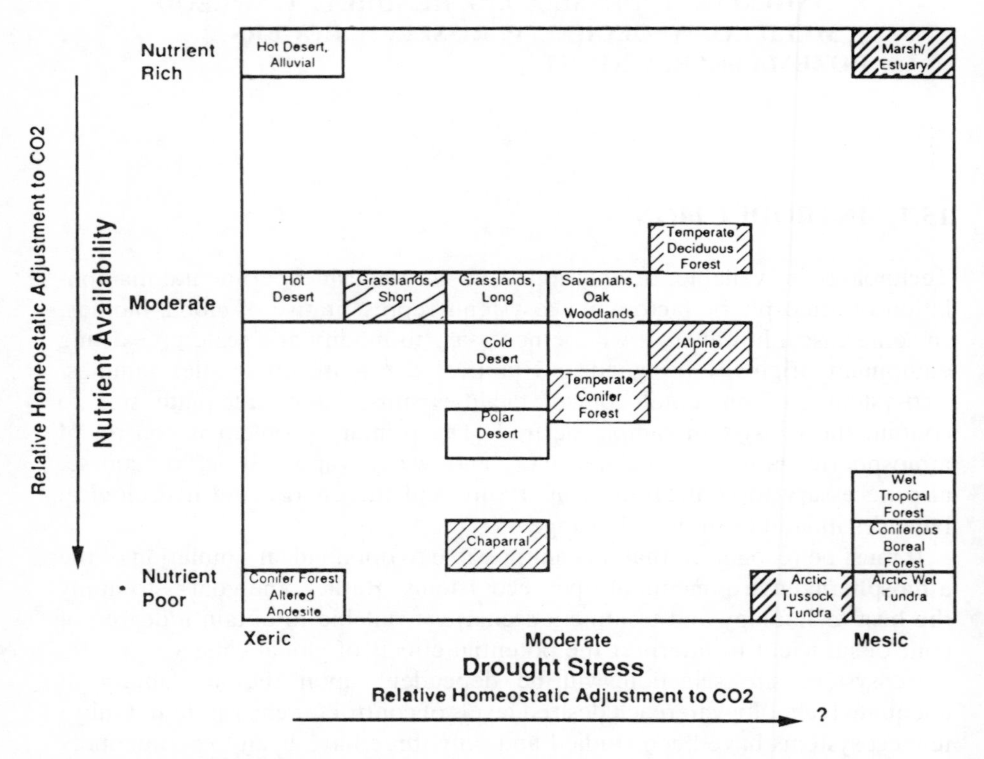

Figure 15.1 Ecosystem types representing the major biomes of the earth arranged on a matrix of increasing gradients of nutrient and drought stress. The largest effects are expected to occur in the most mesic but most nutrient-poor ecosystems. Some ecosystem-level information exists for the marsh/estuary and the Arctic tussock tundra systems. Scattered autecological studies have been completed on the other systems shown in cross-hatching.

boreal forest, (2) tundra, (3) tropical forest, (4) temperate forest, (5) grassland, (6) desert.

Ecosystems will be affected by as much as 11°C mean annual warming; (2) the middle latitudes will have moderate warming but significant drying in the zones of descending atmospheric circulation; and (3) in spite of the prediction that the tropical zone will be least sensitive to weather change, differential species responses to carbon dioxide fertilization will generate important ecosystem modification in the systems of highest biological diversity.

In the past, field researchers examined effects of environmental change using animal enclosures or exclosures, irrigation and fertilization of field plots, automatic or manual rain covers, and simple plastic tents with various degrees of closure. Although these systems have allowed manipulation of some environmental factors (e.g. water, nutrients, herbivores), atmospheric gases could not be controlled. Here we examine various technologies that are available to control atmospheric gases. Current technologies being used for soil and air temperature, water and nutrient modification in ecosystem studies could be applied in the atmospheric control facilities.

Facilities that have been used to control atmospheric gases (e.g. toxic air pollutants, CO_2, water vapor), include leaf chambers, sunlit controlled environment facilities, mobile greenhouses, larger more permanent greenhouses, and open-top chambers. A system of gas lines that deliver CO_2 for release over open fields was described by Allen (1975), and McLeod *et al.* (1983) described a circular plot of stand-pipes for the release of toxic air pollutants. Hendrey and colleagues from the Brookhaven National Laboratory, USA, are refining McLeod's system for the continuous release of CO_2 over an agricultural field. These systems will be described in more detail in the following pages.

15.2 CO_2 MONITORING AND CONTROL SYSTEMS

Experimental facilities that will allow examination of atmospheric CO_2 enrichment and subsequent climate change effects will all have to have the capability of measuring and controlling CO_2 concentration. Various systems were described in Drake *et al.* (1985). Basically, all of the systems have the features shown in Figure 15.2.

15.3 LEAF CUVETTES

The accurate control of atmospheric CO_2 concentration and other environmental factors is possible in small cuvettes. Leaf or branch chambers have been frequently described but here we wish to draw attention to the chamber

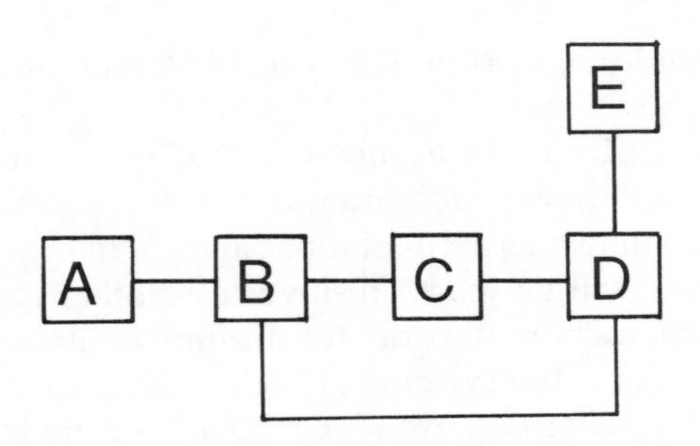

Figure 15.2 Basic schematic for a measurement and control system for CO_2 enrichment experiments in global change research. A = Supply of pure or very high concentration CO_2; B = system of valves, regulators, and flow meters; C = the chamber or area to be CO_2 controlled; D = an appropriate analyzer for the gas of interest with feedback control over the flow control system; E = computerized data acquisition and control programming. If protocol requires subambient CO_2 concentration, unit A will also contain a CO_2 scrubbing unit.

of DeJong *et al.* (1981) because it was specifically designed for laboratory and field work. Materials that are appropriate for field analysis of CO_2 and water vapor, and which have neutral effects on solar irradiance quality, were compared in Bloom *et al.* (1980).

Although these chambers are useful for physiological measurements, they are not appropriate to be used alone for long-term system studies where it is necessary to bring the entire ecosystem into long-term equilibrium. Therefore, we will go on to larger-scale control systems that are appropriate for ecosystem studies.

15.4 OPEN-TOP CHAMBERS

Open-top chambers are essentially plastic enclosures placed around a sample of an ecosystem. Air is drawn into a box by a fan, enriched with CO_2, and blown through the chamber. Air enters near the bottom and flows out the open top. Open-top chambers are inexpensive to build and maintain relative to closed-field chambers. They provide CO_2 control at a fraction of the cost of the FACE method described below.

Open-top chambers were first used for CO_2 enrichment studies by Rogers *et al.* (1983). They used a modification of a design by Heagle *et al.* (1973) for use with toxic gases. Figure 15.3 shows the basic characteristics of the chamber including a frustrum at the top of the chamber. This modification improves the CO_2 concentration profile within the chamber and gives adequate control to within 1 m of the chamber opening.

Drake *et al.* (1989) have reduced the size of the open-top chamber to 1 m in

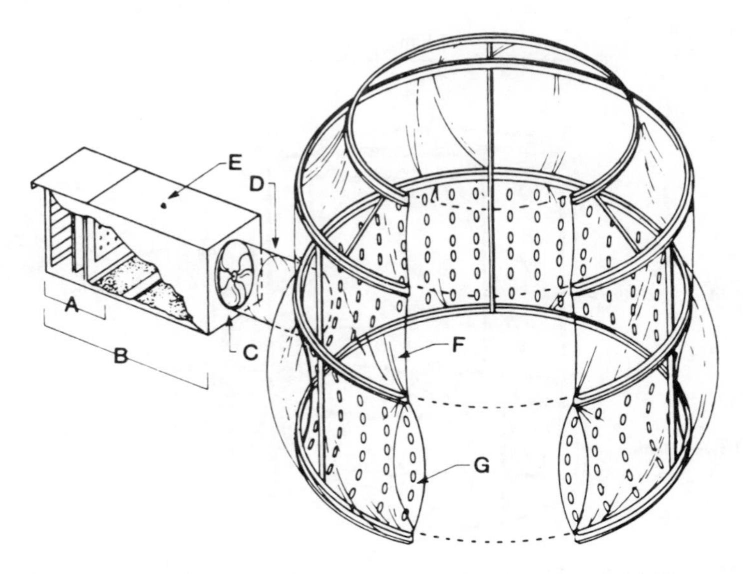

Figure 15.3 Open-top field chamber used by Rogers *et al.* (1983). A = Filters, B = plenum, C = blower, D = air duct, E = port of injection of CO_2, F = single-layer wall, G = double-layer wall with perforations for air entry.

height (Fig. 15.4) for use in an estuarian saltmarsh in Chesapeake Bay, USA.

Open-top chambers have not been used successfully in studies of ecosystems with large vegetation. They are most useful for low-stature systems (e.g. marsh, meadow, tundra, annual grassland). Chambers of 4–7 m in height and 5 m diameter are currently in use for ozone and acid precipitation studies at several research sites of the Southern Commercial Forest Research Cooperative. These studies are autecological on young tree saplings, however, and the 5 m diameter chambers are much too small for studies of complex forest systems containing several species and full-sized forest shrubs and trees.

Open-top chambers are relatively inexpensive to build because they consist simply of an aluminum frame covered by panels of polyvinyl chloride plastic film. Temperature control is affected by air movement. Heagle *et al.* (1979) found less than 1°C differential between inside and outside air temperature. Relative humidity within the open tops is directly related to transpiration rate and is always higher than the external air (Olszyk *et al.*, 1982). Using micrometeorological methods, Wesley *et al.* (1982) found the canopy boundary layer resistance to ozone diffusion to be similar inside and outside of open-top chambers in a soybean field. Chambered and unchambered controls are necessary to establish chamber effect. Many studies (Mandl *et al.*, 1973;

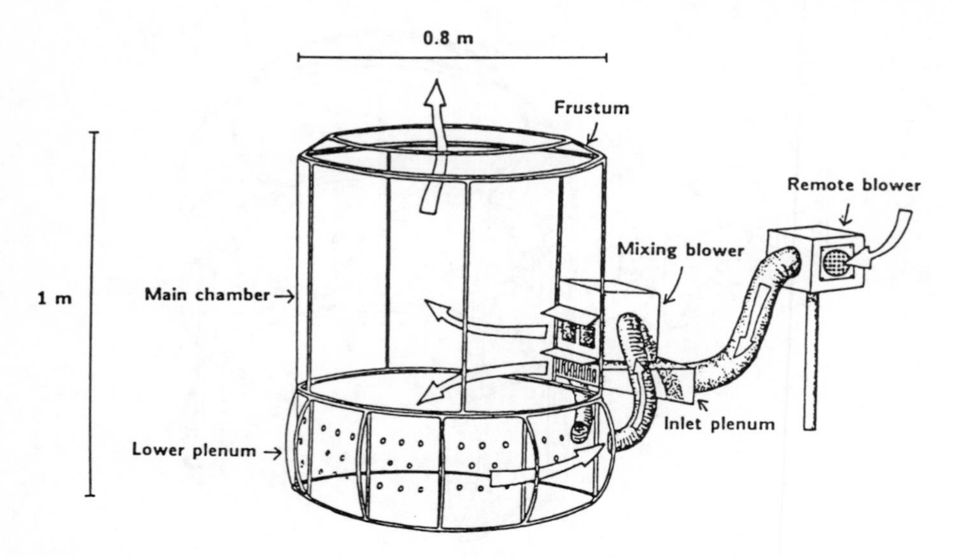

Figure 15.4 Open-top field chamber being used by Drake *et al.* (1989) in a salt marsh. The diagram shows the air path through the chamber and its relative size.

Howell *et al.*, 1979; Heagle *et al.*, 1979; Heggestad, 1980; Rogers *et al.*, 1984a,b, and others) have established the effect of the chambers on environment on the plant response.

15.5 CONTROLLED CHAMBERS (MINI-GREENHOUSES)

Totally closed, transparent chambers or mini-greenhouses are being used to contain samples of ecosystems up to chaparral-sized systems (Paul Miller, personal communication). Most of these attempts, however, have been on crops (Parsons *et al.*, 1980; Jones *et al.*, 1984a) or smaller stature ecosystems such as Arctic tundra (Oechel and Strain, 1985).

Two types of closed systems are portrayed in Figures 15.5 and 15.6. These facilities are designed and operated to provide accurate flexible control of dry-bulb temperature, CO_2 concentration, and vapor pressure of the air. In contrast to the open-top systems, which allow a differential between external and internal variables, these closed systems are designed to maintain small or no differentials. Refrigerated coils, condensing and humidifying systems, and CO_2 adding or scrubbing circuits are used in an attempt to obtain and maintain given levels of environmental control.

Specific methods and equipment for controlling chamber conditions vary but are generally based on: (1) sensors that detect air and leaf temperature,

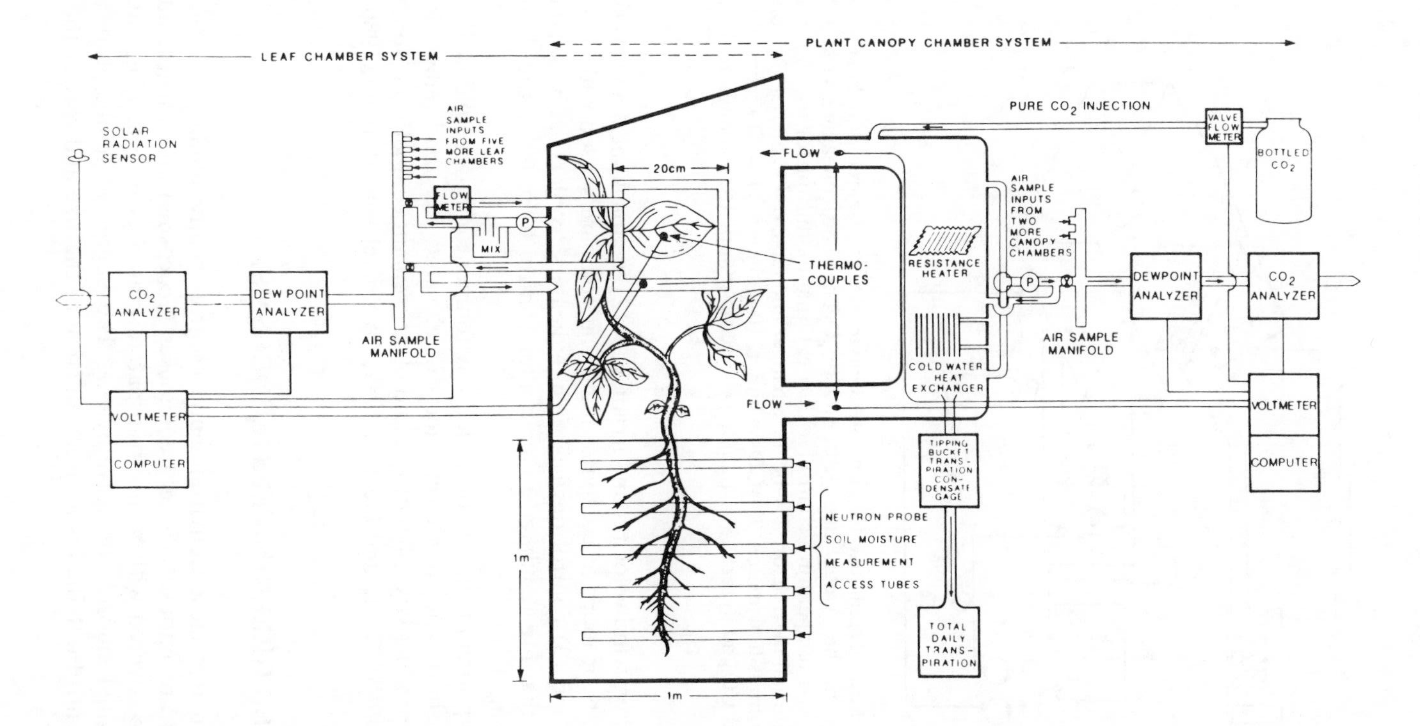

Figure 15.5 Closed system computer-controlled plant environment for CO_2 enrichment study. The plant canopy chamber system was described by Jones *et al*. (1984a). The leaf chamber system was described by Valle *et al*. (1985).

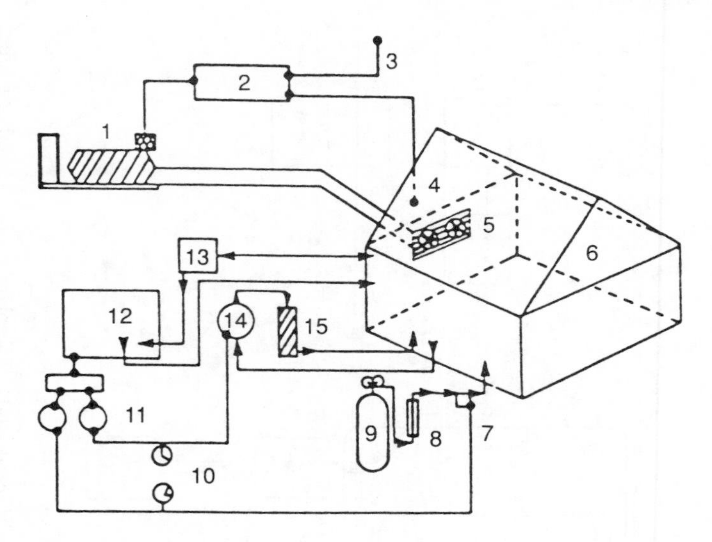

Figure 15.6 Mini-greenhouse for controlled environment research in field studies of global change in the Arctic tussock tundra by Oechel and colleagues (1985) 1 = Compressor, 2 = temperature controller, 3 = external thermistor for use in tracking mode, 4 = internal thermistor, 5 = heat exchanger and fans, 6 = chamber, 7 = on/off solenoid, 8 = mass flow meter, 9 = CO_2 cylinder, 10 = timers, 11 = CO_2 control unit, 12 = infrared CO_2 analyzer, 13 = pump, 14 = scrub pump, 15 = soda lime.

CO_2, humidity, wind velocity, and irradiance; (2) feedback mechanisms such as thermostats or loops in computer logic that compare sensed with desired conditions; and (3) control devices such as heaters, coolers, fans, solenoid valves, etc. that are regulated by computer control to produce the desired treatment conditions.

Recent experiments using controlled environment mini-greenhouses have focused on short- and long-term effects of elevated CO_2 and temperature increase on productivity, phenology, and plant-to-plant competition (Jones *et al.*, 1984b, 1985; Tissue and Oechel, 1987; Overdieck and Reining, 1986).

15.6 FIELD GREENHOUSES (ECOCOSMS)

An ecocosm is a closed environmental control facility similar to a closed refrigerated greenhouse. These greenhouses (ecocosms) are constructed of rigid frames covered with nearly transparent materials such as glass, fiberglass, polyvinyl chloride or polyethylene. The degree of temperature and humidity control obtained is a direct function of the size and cost of the air conditioning and dehumidification system. In ecosytem research these ecocosms must be scaled to the appropriate size and height dimensions to contain a minimal representative sample of the ecosystem of interest.

The purpose of using ecocosms is to allow experimental manipulations of physical and biological variables in studies of ecological interactions through time. Atmospheric gases (CO_2, H_2O vapor, O_2, volatile air pollutants), air and soil temperatures, soil water and nutrients, total solar radiation, precipitation and dry fallout, and wind velocity are controlled and monitored. Additional variables can be controlled in totally closed systems if warranted by the research objective. For example, UV-B can be experimentally manipulated in conjunction with changing CO_2, temperature, and water regime (Teramura and Sullivan, 1989).

Standard greenhouses only partially control environment. Ventilation with outside air is utilized as the primary control procedure. Windows are opened to increase air movement and decrease air temperature. This technique will not work if the objective is to control carbon dioxide level. Consequently, totally closed and controlled greenhouses similar to those of phytotrons (Kramer *et al.*, 1970) are required.

Ecocosms differ from most greenhouses in that they are built over soil and do not have floors. The plants are planted in the ground and grow with more normal root systems, as those shown for a mini-greenhouse in Figure 15.5. This allows the normal development of the rooting zone. Such chambers are currently being installed at the new control facility of the Desert Research Institute of the University of Nevada, Reno.

15.7 FREE AIR CO_2 ENRICHMENT (FACE)

The free air CO_2 enrichment (FACE) methodology for ecosystem research has been viewed by some as a 'real-world' approach. It has been argued that this approach is the best test for the effect of the impending CO_2 enrichment on agricultural and on natural ecosystems. The pros and cons of FACE methodology are presented below at some length because of the uncertainty of the usability of FACE methodology in ecosystem research.

The FACE approach to CO_2 enrichment is to apply a network of vertical vent pipes (McLeod *et al.*, 1983), or other release system (Allen, 1975), near the ground in order to provide elevated CO_2 to the ambient air of the plants. The object is to avoid the need for an enclosure or chamber around the plants. The major differences between FACE and either outdoor controlled mini-greenhouses or ecocosms or open-top chambers are that FACE eliminates the following chamber effects: (1) reduction of the solar radiation, and (2) unnatural wind flow, turbulence, and micrometeorological patterns; further, they potentially permit the study of many ecosystem phenomena.

FACE arrays that have been used include the US Environmental Protection Agency Zonal Air Pollution System (ZAPS) plots with dimensions of 73 m by 85 m for a prairie grassland (Lee and Lewis, 1978). The US Department of Energy used air pollution exposure plots with dimensions of 29 m by 27 m for a soybean crop at Argonne National Laboratory (Miller *et al.*, 1980). The UK Central Electricity Research Laboratories used a circular plot array of 27 m in diameter for a wheat crop exposed to air pollutants (McLeod *et al.*, 1983). This latter system was modified for use over cotton fields for CO_2 release in Mississippi in 1987 and 1988, and in Arizona in 1989–1991 (Hendrey *et al.*, in progress).

Large, uncontained sample areas are an advantage when the system is heterogeneous or diverse vegetatively. Ecological studies of effects of elevated CO_2 on cycles of litter production, organic matter accumulations, soil respiration, nutrient cycling, above-ground competition, and phenology require a large area of uniform exposure and treatment. When replication of sample plots is multiplied by the size and number of plots required to conduct open-area studies, however, logistics and financial requirements may become limiting.

The concentration of CO_2 in a large area supplied through a network of pipes will depend inversely upon wind speed, directly upon the release rate of CO_2 (Allen, 1975; McLeod and Fackrell, 1983) and inversely with vegetation height when mass consistency is taken into account (Hanna *et al.*, 1982). To hold CO_2 concentration constant on the average, the delivery rate must be increased at higher wind speeds, and this requires a feedback mechanism to be included in the FACE design. Nevertheless, it will be very difficult to maintain constant CO_2 under all weather conditions. Under most conditions only the very center of a circular design will have a uniform horizontal distribution of concentration.

The FACE system being used in Arizona by Hendrey and colleagues is composed of an array of 32 individually valved vertical pipes, each containing multiple gas injection ports connected to a 22 m diameter toroidal distribution plenum chamber (Fig. 15.7). A high-volume blower injects air containing variably elevated levels of CO_2 into the plenum torus. The number and location of open vent pipes (Fig. 15.8) is based on both wind direction and speed. The amount of gas metered into the air stream entering the plenum is based on wind speed and real-time measurements of the CO_2 concentration at the center of the array. An empirically derived proportional, integrative, differential control algorithm adjusts the supply of CO_2 and the number of vent pipes releasing gas to maintain the desired concentration within the FACE array.

The gas control system is designed to maintain CO_2 releases that are always upwind of the center of the FACE array. This is achieved by measurement of wind direction and controlling the vertical vent pipe (VVP) ball valves so that

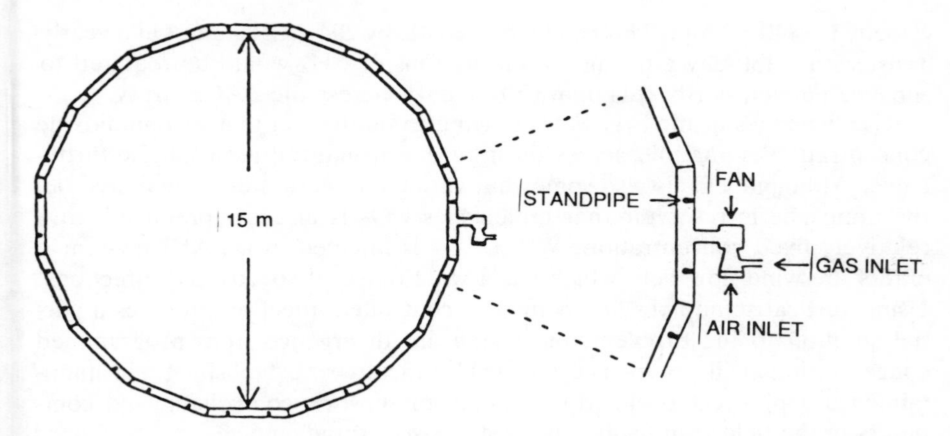

Figure 15.7 Diagram of a FACE ring showing standpipe location and inlet detail for each section of the ring. This diagram represents the 15 m diameter ring used in 1987 and 1988 in Mississippi by Hendrey and colleagues.

the upwind valves are open. At wind velocities lower than a threshold wind speed, all of the VVPs are opened and the system operates without directional control.

It may be possible to provide heating of air in the FACE array to analyze possible effects of climate warming by burning methane and injecting the combustion gas import of the FACE air intake. Starting with the heat of combustion of CH_4, 12 kcal/g, and the specific heat of air, 7 cal/K per mole of air, 0.1036 g of CH_4 will heat 1 m^3 air 4K and will add 144 ppm CO_2. Horizontal air flux through a 22 m diameter, 2 m high FACE array at 1 m/s air

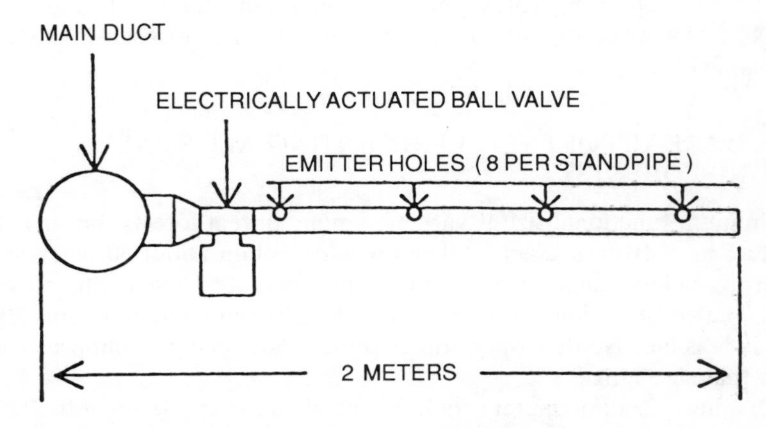

Figure 15.8 Detail of one standpipe on the FACE array described above.

velocity is 2840 m^3/min. This can be heated 4K by 294 g CH_4/min. On a yearly basis, with a 150-day growing season, 64 tons of CH_4 would be required to enhance ambient air by 144 ppm CO_2 and 4K across the FACE array.

Disadvantages of the FACE approach include the fact that carbon dioxide concentration is variable across the ring due to natural atmospheric turbulence. Although CO_2 is also somewhat variable in nature on a smaller scale, the atmospheric reservoir that replenishes CO_2 is already 'premixed' to a relatively fixed concentration. When CO_2 is injected in a FACE system it enters the wind flow field which has a wide range of speeds and directions. Using vertical standpipes and computer-controlled injection provides a partial solution to the problem, but significant divergence from programmed concentrations will occur. The technical difficulty of establishing and maintaining a sophisticated electronic system of sensors, controllers, and computers in the field is difficult. The cost is also a disadvantage.

In spite of the disadvantages described above, FACE may be the preferable technique to use in mid-size to large-stature ecosystems. Open tops have not been applied in these systems and ecocosms have not yet been developed for any system.

15.8 COMMENTS ON INSTALLATION AND ANNUAL OPERATION COSTS OF APPROPRIATE FACILITIES FOR USE IN THREE REPRESENTATIVE BIOMES

This section provides an analysis of the use of the FACE rings, ecocosms, mini-greenhouses, and open-top chambers in three representative biomes. Tundra represents graminoid and low-stature herbaceous perennial ecosystems. Boreal forest represents small to mid-size forested ecosystems, and tropical forest represents forests of large canopy diameter, tall trees. These three systems were considered in this chapter to be the most sensitive systems to global change.

15.8.1 BOREAL FOREST (REPRESENTING ALL SMALL-STATURE FORESTS)

The minimum functional size is variable among boreal forests, but in central Alaska an area 10 m × 20 m (200 m^2) would contain about 30 black spruce trees in the height range of 6–10 m and the associated lichens, mosses, and shrubs. The construction and operation of FACE rings and ecocosms 200 m^2 in size is feasible. Neither open-top chambers nor mini-greenhouses can be built in that size range.

FACE rings (23 m diameter) could be installed for approximately \$137 000 for two rings and \$50 000 for each additional ring. Each ring would consume

6–12 tons of CO_2 per day. Thus, a four-ring array used for a 120-day growing season would require four rings × 6–12 tons CO_2 per day × \$80 per ton of CO_2 × 120 days = \$230 400–460 800 for a 120-day growing season. The range in CO_2 consumed per day is caused by wind velocity. More CO_2 is required in windy ecosystems and on more windy days. Near a natural gas supply it seems feasible to burn methane to obtain CO_2. Electrical generators could be powered with the methane and waste heat could be used for thermal enrichment experiments also. Four rings, each 23 m in diameter, would provide approximately 1600 m^2 of CO_2-enriched space in a graminoid-type ecosystem. Thus, the per m^2 cost of installing and operating an average installation of four rings for 120 days would be \$237 000 for installation + \$230 000–460 000 for CO_2 = \$437 000–697 000/1600 m^2 = \$273–546 per m^2 for year one, and \$230 000–460 000/1600 m^2 = \$144–288 per m^2 for each subsequent year.

All cost estimates in this section are for site installation and logistic operation only. No estimates are given for personnel or additional research costs. No estimates are attempted for maintenance of systems or component replacement. All estimates were made by BRS and are provided for scale comparisons only.

Ecocosms, constructed of infrared-transmitting plastic film stretched over aluminium framing, could be used in the size range of 200 m^2. Height is adjustable with ecocosms of 10 m in height being feasible to construct and maintain. In the boreal forest and the tundra these units could be anchored into the permafrost with stainless-steel rods. Redundant refrigeration systems and fail-safe automatic venting arrangements should be provided. These units would cost about \$250 000 each to construct and would require approximately \$15 000 each for electricity for a 120-day growing season in the boreal forest. Since these systems are closed, with complete internal mixing by fans, the CO_2 costs would be negligible. A computer-based CO_2 monitoring and control system would be required for each complex of ecocosms at a cost of approximately \$40 000 for four 200 m^2 units. The per m^2 cost estimate or four units is \$1 million + \$40 000 + \$15 000/800 m^2 = \$1331 in year one but only \$15 000/800 m^2 = about \$19/m^2 in subsequent years, not counting personnel or research costs.

Mini-greenhouses and open-top chambers could be used for single-tree physiological studies, but are not deemed appropriate for ecosystem-level CO_2-enrichment experiments because a 200 m^2 surface area could not be reasonably controlled with existing technology for these systems.

15.8.2 TUNDRA (REPRESENTING GRAMINOID ECOSYSTEMS)

The minimum functional size for these graminoid-dominated systems is 5 m × 5 m (25 m^2). Height of the vegetation ranges from 0.25 to 2.5 m.

FACE rings for the tundra would require essentially the same installation and per-ring material costs as the boreal forest (i.e. \$237 000 for four rings). Although smaller-diameter rings would probably be employed in the tundra, piping costs are relatively inconsequential. CO_2 costs, on the other hand, would be considerably less; approximately 2 tons/day × 4 rings × \$80 per ton × 90 days = \$57 600 for a three-month season.

Ecocosms can be erected and maintained in graminoid systems. Wind may generate technical difficulties but aluminum framing anchored to the permafrost should be durable. As above, redundant and fail-safe systems will be required to protect the ecosystem samples in case of power loss or catastrophic system failure. The cost of these small ecocosms would be half of that required for boreal forest or approximately \$125 000 per unit. Electricity would cost about \$1600 per unit/three-month season. CO_2 costs would be negligible depending on shipping costs.

Open-top chambers can be used in graminoid systems. Chambers 5 m in diameter would be about 20 m^2 and would contain sufficient area to represent these low-stature but generally closed canopy systems. A nine-chamber complex of 5 m diameter chambers was successfully employed in 1989 and 1990 seasons in tall grass prairie by Clinton Owensby of Kansas State University, and Bert Drake of the Smithsonian Institution has operated small open-tops (1 m^2 diameter) in monospecific estuarian systems for three years.

Mini-greenhouses (including SPAR units) could also be used as demonstrated by Walter Oechel in 1982–85 at Toolik Lake, Alaska, but his units were only 1.6 m^2 × 0.8 m high. His installation costs were about \$10 000 per chamber with an annual operating budget (including some research) of \$300 000 for a 12-chamber complex. CO_2 costs are negligible for all closed systems but electrical consumption is significant. Oechel's system operated 24 hours per day and required a minimum generator size of 10 kW with 20 kW being considerably more stable and dependable.

15.8.3 TROPICAL FORESTS (REPRESENTING ALL LARGE-STATURE FORESTS)

Forests with canopy trees in excess of 20–30 m height generate almost insurmountable technical and financial problems for long-term CO_2 enrichment research. The minimum functional size of rain forest at La Selva, Costa Rica, is estimated to be in excess of 1 ha. A system has to be maintained under CO_2 control for several seasons to obtain equilibrium with the new environment. No ecosystem has yet been enriched long enough to reach carbon equilibrium under high CO_2 levels. In fact, we do not know how long would be required for an ecosystem to reach equilibrium with an enriched atmosphere. Tropical forests would require hundreds of years while annual desert grasslands would presumably require only decades.

Useful physiological information could be obtained by doing single-tree canopies in elevated large open-tops or with elevated FACE rings enriching single trees.

Astrodome-type facilities might be necessary in large-stature forests. The construction costs of such enclosures would be in the millions of dollars ($20–60 million?) with annual operating costs of $2–5 million.

Discussion has been under way for several years, in the US Department of Energy CO_2 program, on the feasibility of using a regeneration approach to the study of tropical rainforest response to global change. Assuming that increasing population and development pressures in tropical areas will generate large tracts of regenerating ecosystems, it would seem to be desirable to study early stages of ecosystem regeneration when plants are becoming established and are small enough to be contained in FACE arrays or ecocosms in the 200–400 m^2 size range. Prices for these units should range from those given for boreal forest to twice those values.

REFERENCES

Allen, L.H. Jr (1975) Line-source carbon dioxide release. III. Prediction by a two dimensional numerical diffusion model. *Boundary-Layer Metero.* **8**, 39–79.

Bloom, A.J., Mooney, H.A., Bjorkman, O. and Berry, J. (1980) Materials and methods for carbon dioxide and water vapor exchange analysis. *Pl. Cell Environ.* **3**, 371–6.

DeJong, T.M., Drake, B.G. and Pearcy, R.W. (1981) Gas exchange responses of Chesapeake Bay tidal marsh species under field and laboratory Conditions. *Oecologia* **52**, 5–11.

Drake, B.G., Rogers, H.H. and Allen, L.H. Jr (1985) Methods of exposing plants to elevated carbon dioxide. In: Strain, B.R. and Cure J.D. (Eds), *Direct Effects of Increasing Carbon Dioxide on Vegetation*, DOE/ER/0238. Nat. Tech. Infor. Ser., Springfield VA 22161.

Drake, B.G., Leadley, P.W., Arp, W.J., Nassing, D. and Curtis, P.S. (1989) An open top chamber for field studies of elevated atmospheric CO_2 concentration on saltmarsh vegetation. *Func. Ecol.* **3**, 363–72.

Hanna, S.R., Briggs, G.A. and Hosker, R.P. Jr (1982) *Handbook on Atmospheric Diffusion*, DOE/TIC-11223. Nat. Tech. Infor. Ser., Springfield VA 22161, pp. 57–8.

Heagle, A.S., Body, D.E. and Heck, W.W. (1973) An open top field chamber to assess the impact of air pollution on plants. *J. Environ. Qual.* **2**, 365–8.

Heagle, A.S., Philbeck, R.B., Rogers, H.H. and Letchworth, M.B. (1979) Dispensing and monitoring ozone in open top field chambers for plant effects studies. *Phytopathology* **69**, 15–20.

Heggestad, H.E., Heagle, A.S., Bennett, J.H. and Kock, E.J. (1980) The effects of photochemical oxidants on the yield of snap beans. *Atmos. Environ.* **14**, 317–26.

Howell, R.K., Kock, E.J. and Rose, L.P. Jr (1979) Field assessment of air pollution-induced soybean yield losses. *Agron. J.* **71**, 285–8.

Jones, P., Allen, L.H. Jr, Jones, J.W., Boote, K.J. and Campbell, W.J. (1984a)

Soybean canopy growth, photosynthesis and transpiration responses to whole season carbon dioxide enrichment. *Agron. J.* **76**, 633–7.

Jones, P., Jones, J.W., Allen, L.H. Jr and Mishoe, J.W. (1984b) Dynamic computer control of closed environment plant growth chambers. Design and verification. *Trans. ASAE* **27**, 879–88.

Jones, P., Allen, L.H. Jr and Jones, J.W. (1985) Responses of soybean canopy photosynthesis and transpiration to temperature in different CO_2 environments. *Agron. J.* **77**, 242–9.

Kramer, P.J., Hellmers, H. and Downs, R.J. (1970) SEPAL: New phytotrons for environmental research. *BioScience* **20**, 1201–8.

Lee, J.J. and Lewis, R.A. (1978) A system for the experimental evaluation of the ecological effects of sulphur dioxide. In: *Proc. 4th Joint Conf. Sensing Environ. Pol.* Am. Chem. Soc., Washington, DC, pp. 40–53.

Mandel, R.H., Weinstein, L.H., McCune, D.C. and Kevemeg, M. (1973) A cylindrical, open top chamber for the exposure of plants to air pollutants in the field. *J. Environ. Qual.* **2**, 371–6.

McLeod, A.R. and Fackrell, J.E. (1983) *A Prototype System for Open-air Fumigation of Agricultural Crops.* 1: *Theoretical design* (TPRD/L/2475/N83). UK Central Elec. Res. Labs.

McLeod, A.R., Alexander, K. and Hatcher, P. (1983) *A Prototype System of Open-air Fumigation of Agricultural Crops.* 2: *Construction and description* (TPRD/L/2475/N83. UK Central Elec. Res. Labs.

Manabe, S. and Wetherald, R.T. (1980) On the distribution of climatic change resulting from an increase in CO_2-content of the atmosphere. *J. Atmos. Sci.* **37**, 99–118.

Miller, J.E., Sprugel, D.G., Muller, R.N., Smith, H.J. and Xerikos, P.H. (1980) Open-air fumigation system for investigating sulphur dioxide effects on crops. *Phytopathology* **70**, 1124–8.

Oechel, W.C. and Strain, B.R. (1985) Native species responses to increased atmospheric carbon dioxide concentration. In: Strain, B.R. and Cure, J.D. (Eds), *Direct Effects of Increasing Carbon Dioxide on Vegetation*. DOE/ER-0238. Nat. Tech. Infor. Ser., Springfield VA 22161.

Olszyk, D.M., Tibbits, T.W. and Hertsberg, W.M. (1980) Environment in open-top field chambers utilized for air pollution studies. *J. Environ. Qual.* **9**, 610–15.

Overdieck, D. and Reining, E. (1986) Effect of atmospheric CO_2 enrichment on perennial ryegrass and white clover competing in managed model-ecosystems. *Oecol. Pl.* **7**, 367–78.

Parsons, J.E., Dunlap, J.L., McKinion, J.M., Phene, C.J. and Baker, D.N. (1980) Microprocessor based data acquisition and controlled software for plant growth chambers (SPAR system). *Trans. ASAE* **23**, 589–95.

Rogers, H.H., Heck, W.W. and Heagle, A.S. (1983) A field technique for the study of plant responses to elevated carbon dioxide concentration. *J. Air Pol. Cont. Assoc.* **33**, 42–4.

Rogers, H.H., Cure, J.D., Thomas, J.F. and Smith, J.M. (1984a) Influence of elevated CO_2 on growth of soybean plants. *Crop Sci.* **24**, 361–6.

Rogers, H.H., Sionit, N., Cure, J.D., Smith, J.M. and Bingham, G.E. (1984b) Influence of elevated carbon dioxide on water relations of soybeans. *Pl. Phys.* **74**, 233–8.

Tissue, D.T. and Oechel, W.C. (1987) Response of *Eriophorum vaginatum* to elevated CO_2 and temperature in the Alaskan tussock tundra. *Ecology* **68**, 401–10.

Teramura, A.H. and Sullivan, J.H. (1989) Interaction between UV-B radiation and enriched CO_2 in three crop species. *Bull. Ecol. Soc. Am.* **70**(2), 278.

Valle, R., Mishoe, J.W., Campbell, W.J., Jones, J.W., and Allen, L.H. (1985) Photosynthetic responses of 'Bragg' soybean leaves adapted to different CO_2 environments. *Crop Sci.* **25**, 333–9.

Wesley, M.L., Egtmon, J.A., Stedmon, D.H. and Yalvao, E.D. (1982) Eddy correlation measurement of NO_2 flux to vegetation and comparison to O_3 flux. *Atmos. Environ.* **16**, 815–20.

Index